Interior Plantscapes

Installation, Maintenance, and Management

Third Edition

GEORGE H. MANAKER

Department of Landscape Architecture
and Horticulture
Temple University

PRENTICE HALL
Upper Saddle River, New Jersey 07458

Library of Congress Cataloging-in-Publication Data

Manaker, George H.
  Interior plantscapes : installation, maintenance,
and management / George H. Manaker. — 3rd ed.
    p. cm.
  Includes bibliographical references and index.
  ISBN 0-13-238494-9 (hardcover)
  1. Interior landscaping.   2. House plants in
interior decoration.   3. Indoor gardening.
  4. Foliage plants.   I. Title.
  SB419.25. M36 1996
  635.9'65—dc20   96-33662
                                                        CIP

Acquisition editor: *Charles Stewart*
Editorial/production supervision: *WordCrafters Editorial
  Services, Inc.*
Cover design: *Miguel Ortiz*
Manufacturing buyer: *Ilene Sanford*
Marketing manager: *Debbie Yarnell*

Printed in the United States of America

ISBN 0-13-238494-9

Prentice-Hall International (UK) Limited, *London*
Prentice-Hall of Australia Pty. Limited, *Sydney*
Prentice-Hall Canada Inc., *Toronto*
Prentice-Hall Hispanoamericana, S.A., *Mexico*
Prentice-Hall of India Private Limited, *New Delhi*
Prentice-Hall of Japan, Inc., *Tokyo*
Simon & Schuster Asia Pte. Ltd., Singapore
Editora Prentice-Hall do Brasil, Ltda., *Rio de Janeiro*

Contents

4   TEMPERATURE   77

5   THE ATMOSPHERE   87

6   PLANTERS   104

7   THE GROWING MEDIUM   124

8 NUTRITION   155

9 MOISTURE   178

10 PROBLEMS   194

✑ Preface ✑

Interior plantscaping is not a fad. It started as part of the back-to-earth, back-to-nature, back-to-the-senses movement of the 1970s and continues unabated. Indoor plants are not nonessential luxuries; they are an integral part of contemporary design, playing an important role in the American way of life. Plants enhance the interiors of our buildings and contribute to the psychological well-being of people. As pollution absorbers, plants improve our general health. Everyone can relate to plants.

The interior landscape profession that evolved over the last 25 years has become legitimized and offers excellent career opportunities. Today, the business has become more complex and competitive, requiring a high level of horticultural knowledge and business savvy.

The first edition of this book was inspired by my inability to find a comprehensive text about the maintenance of foliage plants indoors to use in my classes. Presenting the "hows" and "whys," that book and the revised second edition have been well received by faculty and students in courses relating to interior planting, practitioners in the field, and avid hobbyists.

From simple beginnings a quarter of a century ago, the technology of interior plantscaping has advanced rapidly. This revised third edition updates and expands the previous editions and presents the latest thinking. Changes and additions are based on the latest information available from journals, experiment station and government bulletins, and other publications, as well as personal contacts and observations and participation in professional meetings and seminars.

The basic content is maintained with new or expanded sections relating to such topics as the role of plants in alleviating air pollution indoors, subirrigation planters, and Integrated Pest Management. The list of recommended plants has been revised, the section on flowering plants expanded, and a new topic, novelties, added. The glossary of foliage plant diseases has been expanded, and the source list for planters has been completely revised.

ACKNOWLEDGMENTS

I express my gratitude to Temple University for the resources provided in completing this revision, and to my wife, Veronica, for her encouragement and support. Thanks also to the reviewers, my colleagues, my students, and other users of the book for their comments and suggestions. Acknowledgment is extended to those individuals and companies who provided and gave permission to use photographs, other illustrations, and tables used throughout the book. Finally, my sincere appreciation to the staff at Prentice Hall and WordCrafters Editorial Services for their help in the production of this text.

George H. Manaker

1

Introduction

Foliage plants have been used for decades to add beauty to our building interiors. In the United States, the 1970s marked the beginning of a period of tremendous increase in the use of green plant in homes, apartments, offices, hotels, shopping malls, and other commercial locations. Popularity has never been as great as it is today, with over three-fourths of all American households keeping plants. Architects and interior designers recognize their importance in commercial and public buildings, and design building interiors to accommodate plants.

THE NEED FOR PLANTS

People are dependent on plants, for without the food and oxygen they produce, we could not survive. Drugs, fibers, lumber, oils, and a myriad of other plant products are also important to our daily life. No less significant are the beauty and serenity that plants provide, both indoors and out. Plants and the beauty they provide help to enlarge the imagination and revive the spirit. Ugliness is depressing to those who live or work around it.

Interior planting is not a fad. It started as part of the back-to-earth, back-to-nature, back-to-the-senses movement of the 1970s. Indoor plants are not nonessential luxuries, but an integral part of contemporary design. They are necessities—just as our automobiles and television sets are "necessary," playing a very important role in the American way of life.

The Green Revolution

What has caused this green plant revolution? No single factor is involved. Plants:

1. Fill a psychological need.
2. Enhance our environment.
3. Are a satisfying hobby.

Filling a Psychological Need. In home or apartment, office or shopping mall, plants are an integral part of the decor and satisfy a need. To strip our building interiors of living greenery would be disastrous psychologically and aesthetically.

The human race may have a primal need for plants, with people being inherently unhappy in the absence of greenery. During times when human beings lived in close association with plants, they were part of a natural environmental harmony, but today, Americans are not the outdoor-living people they were 100 years ago, and as much as 90% of our time may be spent indoors. More people are living in apartments in urban environments, isolated from the green of nature.

In today's world, we are constantly bombarded with noise, movement, and visual complexity. Separately and together, they can overwhelm our senses. An environment without plants leads to a psychological deterioration of human life. Because of their simplicity, plants, or natural scenes, reduce physical and mental excitement, cause us to relax, and improve our health. Green is a color of peace and serenity.

Living among plants has a humanizing effect and helps to restore to a healthy condition minds that are burdened with grief, pain, and trouble. Plants exert direct, specific, and positive effects as stimulus objects. When one is under stress, plants—and an opportunity for close association with them—can be very beneficial. The feelings they elicit are not rational; rather, they appeal to the needs of the soul. Everyone can relate to plants.

In planted offices, employees are more content and comfortable, morale is higher, and there is reduced absenteeism. Creativity is enhanced and productivity increased. Standards of dress improve. Office landscaping is definitely here to stay. The federal government has strongly endorsed the use of plants as an integral part of office interiors, and, like thousands of private businesses, utilizes the open-space concept of office design. People should be able to enjoy the space they occupy. Designers must consider the behavioral and psychological needs of people, as well as the physical, and incorporate these into their designs. To do so requires plants. Open-space design, using plants, provides for human needs, allows flexibility in the use of space, and permits changes to be made easily and inexpensively.

Plants satisfy the need to care for living things. A plant is a living organism dependent on the person caring for it and responding to that care. Plants are naturally beautiful and, like people, have their own personality, size, shape, color, texture, and movement. For some, they may substitute for dogs, cats, and other pets. A sense of responsibility is developed as one realizes the degree to which the plant depends upon its owner. The chance to experience success provides a great deal of personal satisfaction, and builds confidence and self-worth. A feeling of accomplishment may also be evident, particularly in those instances in which one has grown an insignificant seedling or cutting into a handsome specimen plant.

Enhancing Our Environment. In recent years, Americans, especially the young, have become keenly aware of their surroundings and the importance of preserving and improving our environment. Interior plantscaping improves the personal environment by making our rooms, work areas, and buildings more aesthetically pleasing and perceptually stimulating. Plants provide warmth and color not otherwise possible, and bring vitality and visual excitement to the plantscaped area. Plants also act as sculpture or living art, thus offering an inexpensive decorating alternative.

Green plants are not only aesthetically pleasing but are functional and aid in directing traffic circulation, providing screens, controlling views and directing the eye, softening harsh architectural surfaces, and giving texture to nondescriptive surfaces. Interior plantscapes reduce large areas to human scale.

In the process of photosynthesis, plants use carbon dioxide from the air and produce oxygen, thereby helping to refresh the atmospheres in our buildings. They may also help cleanse the air by absorbing harmful pollutants, thus helping prevent sick-building syndrome. Indoors sounds from office machines, voices, and ringing phones all contribute to hearing loss and can reduce worker productivity and the quality of life. Green plants abate noise pollution and, with carpets and draperies, act as sound buffers.

Plants and their moist growing media release moisture into the air and increase the relative humidity. Where plants are present, human colds may occur with less frequency and the cracking and drying of wood is reduced.

Indoor plants contribute to our feelings about places and how we perceive the occupants of those spaces. People feel that the quality of occupied space is significantly enhanced by plants.

A Satisfying Hobby. Working among plants provides a means of exercise and relaxation for anyone who enjoys this pastime. Mental and physical well-being are fostered. In addition, indoor gardening provides the opportunity for meeting other people by sharing common interests through membership in any of the numerous plant and horticultural societies throughout the country. As a hobby, plants have helped millions of people of all ages enjoy a more satisfying life and have promoted a fuller interest in the world around them.

Additional Factors. There are several additional factors which have contributed to the green revolution. Personal income is increasing each year, so that more discretionary money is available for the purchase of plants. Foliage plants are readily available in a wide range of varieties and sizes through traditional outlets such as flower shops, garden centers, and nurseries. Less traditional establishments, including supermarkets, department stores, home centers, and discount houses, often have attractive plant displays. Plant care products and decorative accessories are easily acquired, too.

Interest by the media—newspapers, magazines, radio, and television—has helped to educate the public to the use and maintenance of foliage plants indoors, and to spread their acceptance by routinely presenting images in which plants are apparent. The foliage industry itself is involved in efforts to promote green plants and their use for maximum enjoyment. Growers are producing a wide variety of healthy plants to satisfy most needs. Research is providing much-needed information for improving production and maintenance techniques, and plant breeders and explorers are searching for new, improved plants for our indoor gardens.

From all this, a new profession has emerged, that of the interior plantscaper. Experts in the design, installation, and maintenance of interior plantings, these persons provide an essential service to their clients. Much of the content of this text is directed toward these professionals.

No mysterious powers are needed to maintain plants indoors, just an interest in and an understanding of some fundamentals of plant growth requirements.

THE VALUE OF FOLIAGE PLANTS

Since the boom in foliage plants began in the early 1970s, the impact on the ornamental plant industry in the United States has been profound. Between 1971 and 1994, there has been a nearly 1300% increase in the wholesale value of foliage plants (Table 1.1). In 1994, foliage plants accounted for 15.1% of the total value of all floricultural crops, more than the 8.1% reported in 1970, but considerably less than the 30% plus values of the early 1980s. Since 1992, the wholesale value of foliage plants has remained fairly level at 15–16% of the total value of all floricultural crops. The dramatic increase in the wholesale value of foliage plants in 1994 can be accounted for by the recovery of the South Florida production area which was devastated by Hurricane Andrew in August 1992. Florida registered a 34% increase in sales in 1994.

Florida is still the leading state in the production of foliage plants, accounting for 62% in 1994. California is second with about 16%, followed by Texas, Hawaii, Ohio, Pennsylvania, North Carolina, New Jersey, Louisiana, and Michigan.

Although the market for foliage plants has slowed since 1990, it is still expanding faster than the real rate of growth for the general economy. At retail, spending for foliage plants in 1994 was estimated at $1.1 billion, a meager $4.25 per household. Coupled with plant care and related items such as planters, the retail value of the foliage plant market in 1994 exceeded $2 billion in the United States. Supermarkets were the primary source of most plant care products, capturing over $180 million in sales of these items.

THE FUTURE

The future for the foliage plant industry is good, but the boom of the late 1980s and early 1990s has slowed. At the household level, nearly 800 million house plants are

TABLE 1.1   Wholesale Value of Foliage Plants, 1971–1994

Year[a]	Millions of Dollars
1971	37.6
1972	48.1
1973	67.9
1974	111.3
1975	185.0
1976	235.8
1977	271.8
1978	281.9
1979	283.9
1980	312.9
1981	329.1
[b]	[b]
1984	428.6
1985	468.3
1986	521.4
1987	515.3
1988	481.6
1989	488.9
1990	512.1
1991	447.6
1992	416.7
	426.9 (36 states)
1993	417.1
1994	487.0

[a]1971–1976 data from 23 states; 1977–1992, 28 states; 1992 on, 36 states.
[b]Data not available for 1982 and 1983.
Source: Flowers and Foliage Plants, Production and Sales, Sp. Cr. 6–1 (1972–1979), and Floriculture Crops, Sp. Cr. 6–1 (1980–1982, 1985–1995), U.S. Dept. of Agriculture, Washington, D.C.

used in American homes with 82% having at least one. The national average is 10 plants per household, a slight decline from 1990. A lack of new and different indoor plants and a shift to outdoor gardening may have contributed to the reduction. Families with incomes exceeding $40,000 tend more indoor plants than those with lower incomes. In general, higher incomes translate into more plants. Heads-of-households under age 30 have fewer plants, averaging eight, than their older counterparts, who average 12. One-person households have the fewest plants, while households with no children living at home have the most. Sixty-one million of America's 106 million homes account for two-thirds of all the house plants used in the United States.

The population of the United States is increasing and exceeded 260 million people in 1994, with per capita income exceeding $21,000. Baby boomers are aging and their buying power is increasing. Their attitudes have changed; they are no longer obsessed with the quantity of possessions and the symbols of "belonging" (expensive cars, for example). Their current focus is on the quality of life including concern for the environment. Plants address this interest. Baby boomers represent a significant market for foliage plants.

The number of households increases each year and totaled about 106 million in 1990. Although family units are smaller, there is still excellent potential for foliage plant sales.

Quality is essential; consumers today are quality- and value-conscious. They are becoming aware of the benefits of plants indoors, and are selecting plants that will make long-term contributions to the beauty and health of home or office. With a greater understanding of plants and their care, retail consumers are seeking plants that tolerate less water, low light, and the lower temperatures of our building interiors, and that resist pests. Flowering plant are also in demand. Consumers will buy from knowledgeable specialists who offer a wide selection of quality plants at competitive prices.

For the interiorscape industry, important changes are occurring. Business is more complex and competitive than ever with the potential for successful firms to have a fantastic future. Over the past 20 years, several large enterprises have emerged to dominate the industry. Formed from mergers (acquisitions) and the opening of branches, they represent a significant presence in large markets. Rentokil, Inc., Tropical Plant Services, is the industry giant with over $83 million in sales in 1994. It has grown to national status through acquisitions since its founding in 1963. Plantscaping by Orkin, established in 1990, was number two in 1994 with sales of $18.5 million. Other familiar corporate names with interiorscape divisions include TruGreen-Chemlawn and ServiceMaster. One large firm, Foliage Design Systems, is an international franchise. Numerous other multimillion dollar businesses operate regionally around the country. Large firms are committed to growth and have the resources to do so.

Small firms continue to have an important place in the interior landscape industry with more than 2,600 such businesses in operation in 1994. Concern about the future exists among the owners of small firms. Competition with large firms in major markets will be healthy. There's enough business for everyone. To be successful, firms must find and fill a market niche, provide high-quality, personal service, and demonstrate sincere concern for, and commitment to, their clients. Opportunities also exist in small cities and towns where there is room to grow. Consideration should be given to the establishment of a new interiorscape business, or relocation, in less urban areas.

The recent business slowdown has caused many interiorscape firms to diversify. Among related services offered by some companies are production of special events such as grand openings, decorations for parties and conventions, and displays for holidays other than Christmas. Some firms have established retail foliage and flower shops, a return to the origins of many of the first interiorscape businesses.

Government regulations in the future will become increasingly burdensome. Every business will need someone to remain abreast of federal, state, and local regulations and to assure compliance. Consider, for example, licensing, certification, right-to-know laws, the Americans With Disabilities Act, the Occupational Safety and Health Administration (OSHA), the Federal Insecticide, Fungicide, and Rodenticide Act (FIFRA) and related state regulations, and the EPA Worker Protection Standards, to name a few.

A direct correlation exists between construction and interiorscaping, with new space generating new plantings. After a slowdown in building in the late 1980s and early 1990s, signs of improvement are appearing. In 1995, the supply of commercial space exceeded demand by a wide margin, a trend that is likely to continue; however, construction of healthcare facilities is increasing as is new building and renovation at two- and four-year educational institutions. Both offer opportunities for interior landscaping. Retail construction has been characterized by renovation and expansion rather than new construction. These changes can mean updated and/or upgraded interior plantscapes.

Professional plantscapers are looking toward continued prosperity, although they anticipate a more competitive business environment in which price and quality will be major client considerations. Expressed needs of plantscapers include standardized products, techniques to maintain healthy plants cost-effectively in environments designed for human comfort, safe chemicals, new plants tolerant of low light and humidity and temperature variation, and improved shipping and handing techniques.

Some say that even though interiorscaping has become legitimized with highly experienced and professional practitioners, the professional image needs further improvement. Large firms have the financial means to create a positive image: full-color brochures, well-dressed, articulate sales people, and business knowledge, for example. Many smaller businesses do not; however, personal computers have helped many project an improved image, expand their client base, and improve their profitability.

Although significant improvement has been made, more cooperation among those concerned with the building's interior—architects, interior designers, plantscapers, and growers—is desired. Educational programs to keep up to date on the latest technical and business information are another essential for continued success.

At the commercial level, plants are here to stay. Garden spaces in offices, restaurants, banks, hotels, and shopping malls create a company image and provide a comfortable environment. Employee morale is improved. Administrators are aware of the role plants play in a successful business and are not likely to strip their buildings of greenery.

The foliage plant industry, too, is promoting the product on an industry-wide basis and helping to educate the public to the importance of plants to daily life. The future seems bright indeed for the industry and for knowledgeable, qualified specialists in interior plantscaping.

KINDS OF PLANTS USED

Tropical and subtropical herbaceous and woody foliage plants are by far the most important and predominant group used in interior landscaping, either at home or commercially. Flowering pot plants and bulbs are often incorporated into the landscape, especially for seasonal effects. Pot mums, poinsettias, tulips, and daffodils are suited to this purpose but, since they are relatively short-lived, must be changed before they begin to fade. Less frequently used indoors are flowering and fruiting shrubs. Cacti and succulents are excellent plants for interior use either at home or commercially. At home, vegetables and herbs make interesting plants for growing indoors, perhaps in the kitchen.

WHERE TO USE PLANTS

If the environmental requirements are satisfied, there is practically no place where living, green plants may not be used. At home, any room, including the basement and garage, may be satisfactory. Plants may be used on tables, bookcases, and desks, on floors or pedestals, hanging in windows and on windowsills, or as room dividers. They may be single specimen plants or attractive groupings, enhancing the decor and creating an atmosphere of warmth and beauty (Figures 1.1 to 1.4).

In commercial and public buildings, extensive plantings have become an integral part of the design and decor (Figures 1.5 to 1.9). Consider the interior plant-scaping of shopping malls, banks, hotel lobbies, restaurants, and offices. Ranging from a few pots on desks or files, to space dividers and large specimen plants, to mas-

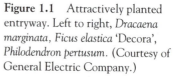

Figure 1.1 Attractively planted entryway. Left to right, *Dracaena marginata, Ficus elastica* 'Decora', *Philodendron pertusum*. (Courtesy of General Electric Company.)

Figure 1.2 A windowsill garden offers the opportunity to display a wide variety of plants.

Figure 1.3 A small table plant enhances the decor of the room: *Syngonium podophyllum*.

Figure 1.4 Individual specimen plant: *Araucaria heterophylla*. (Courtesy of Everett Conklin.)

(a)

(b)   (c)

Figure 1.5 Plants in shopping malls: (a) *Brassaia actinophylla* underplanted
with *Dracaena deremensis* 'Janet Craig' (on the far left) and *Ficus benjamina*
underplanted with *Dracaena deremensis* 'Warneckii' (second from the left); (b)
Alexander palm, *Ptychosperma elegans*; (c) individual accent plant
Dieffenbachia maculata.

Figure 1.6 Atrium lobby of commercial building: *Phyllostachys nigra*, black bamboo, and seasonal floral plantings.

(a)   (b)

Figure 1.7 Plants used as dividers: (a) individual planters with *Dracaena deremensis* 'Janet Craig' separate the seating area from the rest of the room; (b) raised divider with *Dracaena deremensis* 'Warneckii' and *Chamaedorea elegans*.

Figure 1.8   File area: *Cissus rhombifolia*.

Figure 1.9   Atrium of an office complex: *Washingtonia robusta*.

sive gardens such as those at the Opryland Hotel in Nashville and Minnesota's Mall of America, green plants provide a pleasant environment in which to work, shop, or play, and a warm welcome to all who visit.

SUMMARY

The popularity of foliage plants in the United States has never been as great as it is today. Green plants may no longer be a nonessential luxury, but a necessity to the American way of life, filling a deep psychological need, enhancing our environment, and providing leisure-time activity. Increased discretionary income, wide availability, new varieties, and better quality have also contributed to a boom in green plants.

With a vast increase since 1971, the foliage market accounts for 15.1% of the value of all floriculture crops and is showing signs of stabilizing. The future seems good, with consumers demanding new varieties, larger sizes, and high quality. The anticipated increase in population, number of households, and incomes over the next several years all bode well for the green plant industry. In commercial buildings, green plants are here to stay. They create a company image and provide a psychologically comfortable environment with improved employee morale.

Tropical and subtropical green plants are the mainstay of interior planting, although they are frequently used in conjunction with flowering plants and cacti and succulents. In homes and commercial and public buildings, there is practically no place in which plants cannot be used if the environmental requirements prevail. Great rewards exist for the foliage plant hobbyist, and knowledgeable, qualified interior plantscapers should find rewarding careers.

REFERENCES

"An Unlikely Industry Barometer Tracks Trends," *Interiorscape* 14(2): 22–23, 1995.

BLACK, B.: "Horticulture Therapy Comes of Age," *Garden Journal* 21(1): 8–11, 1971.

CAREW, J.: "The Composition of Horticulturists," *Proceedings of the 17th International Horticultural Congress* 2: 89–95, 1966.

CONKLIN, E.: "Indoor Landscaping," *Horticulture* 50(3): 48–52, 1972.

CROCKETT, J. U.: "What Can a Plant Do for You?" *Horticulture* 55(3): 4–5, 1977.

DELL, J.: "The Art of Commercial and Residential Plantscaping," *Interscape* 6(5): 22, 24, 1984.

FERGUSON, B.: "House Plants as Decoration," *Plants and Gardens* 28(3): 31–37, 1972.

FIELDS, S. F.: "Plant Trees and Flowers in the Living Room of Your Apartment," *Florists' Review* 147(3821): 26ff., 1971.

Floriculture Crops, 1985–1994 Summaries, Sp. Cr. 6–1 (86–95), U.S. Dept. of Agriculture, Washington, D.C., 1986–1995.

Floriculture Crops, Production Area and Sales, 1978 and 1979, Intentions for 1980, Sp. Cr. 6–1(80), U.S. Dept. of Agriculture, Washington, D.C., 1980.

____, *1979 and 1980, Intentions for 1981*, Sp. Cr. 6–1(81), U.S. Dept. of Agriculture, Washington, D.C., 1981.

____, *1980 and 1981, Intentions for 1982*, Sp. Cr. 6–1(82), U.S. Dept. of Agriculture, Washington, D.C. 1982.

____, *1984 Summary, Intentions for 1985*, Sp. Cr. 6–1(85), U.S. Dept. of Agriculture, Washington, D.C. 1985.

Flowers and Foliage Plants, Production and Sales, 1970 and 1971, Intentions for 1972, Sp. Cr. 6–1(72), U.S. Dept. of Agriculture, Washington, D.C., 1972.

____, *1971 and 1972, Intentions for 1973*, Sp. Cr. 6–1(73), U.S. Dept. of Agriculture, Washington, D.C., 1973.

____, *1972 and 1973, Intentions for 1974*, Sp. Cr. 6–1(74), U.S. Dept. of Agriculture, Washington, D.C., 1974.

____, *1973 and 1974, Intentions for 1975*, Sp. Cr. 6–1(75), U.S. Dept. of Agriculture, Washington, D.C., 1975.

____, *1974 and 1975, Intentions for 1976*, Sp. Cr. 6–1(76), U.S. Dept. of Agriculture, Washington, D.C., 1976.

____, *1975 and 1976, Intentions for 1977*, Sp. Cr. 6–1(77), U.S. Dept. of Agriculture, Washington, D.C., 1977.

____, *1976 and 1977, Intentions for 1978*, Sp. Cr. 6–1(78), U.S. Dept. of Agriculture, Washington, D.C., 1978.

____, *1977 and 1978, Intentions for 1979*, Sp. Cr. 6–1(79), U.S. Dept. of Agriculture, Washington, D.C., 1979.

GAINES, R. L.: *Interior Plantscaping*, Architectural Record Books, New York, 1977.

HAMMER, N.: "Here's Looking at You . . .," *Interiorscape* 4(2): 6–13, 1995.

HARRIS, R.: "The Power of Plants," *Interior Landscape* 11(4): 26–30, 1994.

LEWIS, C. A.: *Healing the Urban Environment: A Person/Plant Viewpoint*, The B. Y. Morrison Memorial Lecture, American Institute of Planners, 1978.

____: "People–Plant Interaction," *American Horticulturist* 52(2): 19–24, 1973.

LOHR, V. I.: "Quantifying the Intangible," *Interior Landscape* 9(8): 32–39, 1992.

MCDUFFIE, R. F.: "Putting Plants to Work," *Interior Landscape Industry* 1(7): 40–43, 1984.

MENNINGER, W., and J. F. PLATT: "The Therapy of Gardening," *Popular Gardening* 8(6): 54, 1957.

MOREY, J. and C. MOREY: "'95 Interiorscape Contractor 25," *Interiorscape* 14(6): 29–31, 1995.

OTT, R.: "The Crystal Ball," *Interior Landscape* 10(4): 44–48, 50–51, 1994.

"Potted Foliage Market Continues Downward Trend, Construction Markets Offer Some Promise," *Interior Landscape* 11(1): 7, 1994.

RELF, P. D.: "Rising Demand for Horticultural Products," Lecture, Session #C15, Agricultural Outlook Conference, Washington, D.C., 1990.

_____: "The Use of Horticulture in Vocational Rehabilitation," *National Council for Therapy and Rehabilitation through Horticulture Newsletter* 9(6): 2–3, 1982.

SMITH, I.: "Consumer Statistics: Who Gardens? Why?," *Grower Talks* 59(2): 66–72, 1995.

TALBOTT, J. A., D. STERN, J. ROSS, and C. GILLEN: "Flowering Plants as a Therapeutic/Environmental Agent in a Psychiatric Hospital," *HortScience* 11(4): 365–366, 1976.

"The Value of Plantscaping," *Interscape* 4(23): 24, 1982.

History and Origins

HISTORY

Plants have been grown in containers since early times, but their extensive use indoors is recent. The first evidence of plants indoors is difficult to ascertain, but they were probably used in homes in China 3,000 years ago. The ruins of Pompeii indicate the use of interior plants 2,000 years ago. Large homes in Pompeii had plants in terra-cotta pots in atria or peristylia, while smaller houses contained a few pots in a light well. The identity of the plants is difficult to determine, but probably included laurels, myrtles, lemon or citron, and perhaps evergreen shrubs and exotic plants.

Gardening originated over 3,500 years ago in the Euphrates Valley, where people first began to live in settled communities. Early crops were edible and later medicinal. Stone carvings of the period show that plants, especially fruit trees, were grown in containers. Ornamental gardening evolved later and was the product of urban civilization. The Hanging Gardens of Babylon (814–810 B.C.), rebuilt in 605 B.C. by Nebuchadnezzar, is believed to have been located near present-day Baghdad, Iraq. It is the best classic example of growing plants in water-proofed, stone vessels, on terraces, and using an irrigation system.

Egypt. Egyptians loved flowers and green plants, and used them abundantly to adorn monuments and as presents to the gods. Tomb paintings reveal that some earthenware containers were used for growing small trees and shrubs and were set about the garden. The homes were strikingly bare, and no plants were used indoors until the third century B.C., when plants in terra-cotta pots became adornments for the inner courts.

Greece. The origin of true pot gardening may have been in Greece, where Greek women grew lettuce, fennel, wheat, and barley in pots for the festival of Adonis, the spirit of nature and plant life. Plants were displayed on housetops and decorated statues of Adonis.

Celebration of the festival, which was in midsummer, was borrowed from the Babylonians, Assyrians, and Phoenicians. Adonis was believed to die in the autumn and descend into the underworld, to be rescued by Aphrodite and brought back to earth in the spring. Soon, other plants were used for the decoration of courtyards and rooftops. In the days of Theophrastus (late fourth to early third century B.C.), pot gardening was well established, as evidenced by pot fragments dug up in 1936, which date to that period. Plants were used for all-season decoration, not just for the Adonis festival. Roof gardens were common. From Greece, the practice of container and roof gardening spread to other areas of the Mediterranean.

Rome.   Scrolls of Virgil, Horace, and Seneca indicate that the custom of the Greeks was copied by the Romans, who used stone containers for growing ornamental plants, particularly in the atria and peristylia of their homes. The house and garden were integral. Roman atria were open to the air so that natural light was plentiful, and they provided sites for the first use of plants within architecturally confined spaces. As the focal point of the home, atria frequently housed large numbers of plants (Figure 2.1). Potted plants were used on roofs.

The Adonis rite gave rise to potted plants along terraces and patios and around pools. Windowsill pot gardens were common and window boxes were used for the

Figure 2.1   Re-creation of a peristyle garden of an ancient Roman country house.

first time. Information on the types of plants is not easily available, but citrus was apparently a frequently used species, in addition to rare flowers and tropical plants.

By the time of Pliny (first century A.D.), gardening in Rome was at a pinnacle of development. Hothouses and conservatories with windows of talc or mica were built. Glass was used after A.D. 290. The structures were heated and used for growing such plants as roses, grapes, and lilies.

Dark Ages.   With the invasion of Rome in the fifth century, ornamental garden-ing was set back. For the next 1,100 years, gardening was preserved by the religious, Christians and Muslims.

Medieval gardens were functional, producing medicinal and culinary herbs. Some earthenware was used. During the Romanesque and Gothic periods, monas-teries grew fruit, vegetables, and herbs.

In the eleventh and twelfth centuries, gardens grew within castle walls. Pot-ted plants were used on walls, in flower beds, and on turfed benches for ornamental effect. A great many new plants were introduced, especially in Europe, as a result of the Crusades (A.D. 1100–1300).

The Arab influence, particularly in Spain (A.D. 712–1492), left a tradition of reflecting pools and plants in terra-cotta pots on patios. These were the only deco-rative forms of pot culture prior to the eleventh century.

By the end of the Dark Ages, an appreciation for decorative plants began to evolve, and flowers were often grown in pots of earthenware or metal.

Renaissance.   The Renaissance marked a revival in pot gardening. The walled gar-den of potted plants designed by Alberti for Giovanni Rucellae toward the end of the fourteenth century attests to this fact. Orangeries were a common feature on the estates of the wealthy, as were greenhouses in later years. Citrus in pots and boxes were used outdoors in summer and moved into the shelter of orangeries for the win-ter. The Palace of Versailles (1683), for example, had an orangery that could hold 1,200 oranges and 300 other tender shrubs (Figure 2.2). Built in 1684–1686, the orangery consists of three vaulted galleries 42 ft high and 42 ft deep. The longest is 168 yds long. Sunken on the north and open on the south, the structure is not heat-ed. Winter temperatures never go below 43°F.

The period was one of great exploration, as marked by the discovery of Ameri-ca in 1492, India in 1498, and Java in 1511. As a result, new plants were introduced into southern Europe, enhancing the collections of the wealthy. So great was the interest that the first botanical garden was established at Padua in 1545. In 1599, the first artificially heated greenhouse was built at the botanic garden in Leiden, Hol-land. By 1633, plants of such present-day genera as *Asparagus, Coleus, Crassula, Hed-era, Hyacinthus, Narcissus,* and *Saxifraga,* as well as numerous ferns, were common.

In Germany during the sixteenth century, decorative plants in clay pots were used on windowsills. Many consider this the beginning of the pot plant habit by the common man.

During the seventeenth century, particularly in England, greenhouses on the estates of the wealthy housed oranges. A greenhouse at that time meant a building

Figure 2.2   Orangery at the Palace of Versailles.

used to protect tender plants against the cold, and was maintained at a low temperature. William and Mary (late seventeenth century) were keenly interested in horticulture. They were collectors of exotic plants and maintained three hothouses for their display. Lead and stone containers became popular in the period. In Italy, flowering plants in pots were used in the loggia of villas. In America, settlers had little time for plants indoors.

Eighteenth Century.   During the eighteenth century, great interest was shown in exotic plants and the introduction of new plant material. By the middle of the period, over 5,000 species had been introduced. Improvements in glass manufacture after 1700 resulted in flat, relatively clear panes which transmitted more light. On pretentious estates in England, orangeries gave way to conservatories and hothouses, where ideal temperatures could be maintained year-round.

There was little interest in exotic plants in America in the eighteenth century, except by the aristocracy. The first American hothouse was built in New England in 1737 by Andrew Faneuil. It housed tropical plants including banana, pineapple, and mimosa.

Nineteenth Century.   As the nineteenth century began, tropical and subtropical plants were used as decorative accessories in hothouses and conservatories and thus remained an interest of the aristocracy, for they had the time and the money. Others had enough daily contact with plants, and did not want them indoors as a hobby. As early as 1816, foliage plants were used indoors, as shown in the paintings of the period. By the end of the first quarter of the nineteenth century, conservatories with rare and exotic plants became the true mark of elegance and prestige among the ris-

TABLE 2.1   Plants Listed in 1814 Which are Popular Today

Adiantum	Cactus	Nerium
Agave	Camellia	Pelargonium
Aloe	Ceropegia	Phoenix
Alpinia	Chamaerops	Pittosporum
Aralia (Polyscias)	Chlorophytum	Plectranthus
Areca	Cissus	Pothos
(Chrysalidocarpus)	Citrus	(Epipremnum)
Asparagus	Coffea	Pteris
Azalea	Crassula	Sansevieria
(Rhododendron)	Cycas	Sedum
Begonia	Dracaena	Stapelia
Bryophyllum	Ficus	Tradescantia
(Kalanchoe)	Heliconia	

ing middle class, and many popular genera of the period are still recommended (Table 2.1).

At first, indoor gardening was more popular in Europe than in America. In England, nearly everyone had a window garden or greenhouse, and scarcely any family of intelligence did not know something of the culture and propagation of indoor plants. By midcentury, mansions of the wealthy and homes of the middle class overflowed with potted plants, hanging baskets, and wrought-iron stands holding tiers of plants. Public conservatories became more common (Figure 2.3).

The Industrial Revolution had a strong influence on the development of

Figure 2.3   Palm house at Kew Gardens, constructed 1844–1848.

indoor gardening, in that it created an urban middle class with leisure and money for plants. Initially a refined, feminine activity for the rich, the pastime attracted the interest of the poor, the rural communities, and men, and was encouraged as morally stimulating. The practice of horticulture was considered a mark of distinction and rapidly became a respected hobby and art for all classes. Gardeners were considered craftsmen.

Window gardening and plant decoration indoors became universal in America and Europe after 1850, and may have reflected the materialism of the nineteenth century. The presence of flowers and plants indoors was considered essential to the development of refinement and taste, and was one of the most elegant, most satisfactory, and least expensive ways of showing grace. Not only did plants make the home attractive, but they contributed to one's well-being by bringing joy, happiness, and peace. They eased pain and sorrow, cheered the sick, and provided rest for the weary. Indoor gardening was an excellent hobby, especially for invalids. In addition, the culture of plants indoors provided opportunities for the introduction of the study of natural history. Parlor gardening was open to everyone, and impossible for none.

The cultivation of ferns during this time period is of special note. Prior to 1830, there was little interest in growing ferns indoors. Although new species were brought back by explorers and the method of growing spores had been perfected, few people found ferns attractive. In 1831, the introduction and subsequent development of glass-enclosed growing cases by N. B. Ward were important to the rise in the popularity of ferns. Called Wardian cases, the boxes provided an ideal environment for small ferns and other plants, both flowering and foliage, and may be considered the start of terrariums. Ward grew 30 species in cases by 1833, and by the 1840s, Wardian, or fern, cases were fashionable in large drawing rooms in Europe. They did not become popular in America until the Victorian period. Modifications included bell jars and stone vases with glass frames (Figure 2.4). Many books were published, 14 in 1854–1855 alone. By 1860, there were no fewer than 818 species of fern in cultivation. The fern mania lasted until the end of the 1860s. No longer a status symbol, ferns and fern cases were used by everyone and interest in them waned.

Victorian Period. Indoor gardening was a popular, elegant, and harmless pleasure in England and North America during the Victorian period of the late nineteenth century. Plants added cheerfulness to a room and gave pleasure to those who cared for them.

Oversized tropical plants were massed in homes, hotels, theaters, and other public places, or placed in conservatories or enclosed window gardens resembling conservatories (Figure 2.5). The conservatories were often attached to the house, with plants planted permanently in beds or borders to display their natural characteristics. Winter- and spring-flowering plants were especially popular in conservatory plantings.

Most nineteenth-century homes had plants in the parlor or in window boxes or gardens. The number of plants that could survive gas lighting and coal heating was limited. Homes were cold and dark, and dust was a problem. Blinds, shades, or

Figure 2.4 Victorian fern case. Introduced in 1831 by N. B. Ward, they were the first terrariums.

Figure 2.5 Victorian parlor.

thick curtains used to darken rooms to prevent fading of fabrics and wood reduced the light intensity for plant growth. Ventilation was often poor. High humidity from heating stoves was an asset, however.

Foliage plants were popular for parlor decoration, and the most commonly used genera today were introduced in the nineteenth century. *Dracaenas* were considered the plants best suited for indoor use and had wide popularity. *Cordyline (Dracaena) terminalis* was the most frequently used plant. Palms were considered the most useful and durable plants and were used extensively, as were rubber plants, ferns, cycads, crotons, *Begonia*, and *Aspidistra*. Other important plants included bromeliads, *Fittonia, Maranta, Pellionia*, pineapple, orange, lemon, *Hoya carnosa, Beaucarnea*, cactus, *Ixora, Stapelia, Anthurium, Camellia, Yucca, Hedera, Agave, Acacia, Aloe, Aucuba, Alglaonema, Cissus, Araucaria, Caladium, Dieffenbachia, Aralia (Polyscias), Bambusa, Musa, Philodendron pertusum (Monstera deliciosa), Coleus*, and *Pittosporum*. Many flowering plants were recommended, including rose, *Fuchsia*, heliotrope, geranium, *Lantana, Tropaeolum, Ipomea*, carnation, *Azalea, Petunia*, and *Verbena*. Potted bulbs were another frequent component of the window garden.

Hanging baskets were important during the Victorian era, and wire, earthenware, and zinc-lined planters containing ivy, ivy-leaved geraniums, *Saxifraga, Tradescantia*, ferns, and foliage plants were common.

Porous pots in 10 sizes, ranging from 2 to 18 in. were commonly used. Because they were not very attractive, these growing containers were usually inserted into decorative containers made of ceramic, iron, brass, enameled tin, wood, or stone. The plants were arranged throughout the room, enhancing the clutter of Victorian houses. Suitable locations for plants included mantle, étagère, plant stand, or floor. Trellises were even constructed around windows to accommodate vines. Fern cases in various sizes and shapes mounted on ornamental urns and pedestals fit right into the decorative scheme of Victorian homes.

Management was not easy, and many of the recommendations for plant maintenance made during the nineteenth century are similar to those of today. According to LaMance (1893), patience, careful selection of plants, and a knowledge of what and what not to do were required. Success depended not on one factor, but several. No luck was involved, but careful thought.

Starting with quality plants carefully selected was essential. Consideration should be given to the space available, and the number and kind of plants needed. Were they to be large specimens, small plants, or a combination? Plants should be chosen to please taste, to suit the room, and to relate to the space, heat, and light available. As many plants should be kept as could be cared for well and brought to the highest standard of beauty.

Plenty of light and warmth during the day, with darkness and cooler temperatures at night, were the standard recommendations. East-facing windows receiving morning sun and bay windows were considered ideal. If not available, the sunniest windows were recommended. Shady windows were suitable for Wardian cases and for all foliage plants except those with highly colored leaves. Most flowering plants would not bloom in shady windows and were not recommended. Plants were to be turned daily to keep them straight.

A temperature of 60 to 70°F during the day and not less than 40 to 45°F at night was recommended. Sudden, extreme fluctuations were to be avoided. To protect plants from the cold, dropping curtains over them or pinning newspaper around them was suggested. Choosing cold-tolerant plants was also desirable. Overheated rooms (70 to 80°F) caused problems from both the heat and the low humidity. Selection of heat-tolerant plants, ventilation, and promotion of atmospheric moisture were recommended. LaMance stressed the importance of insulating walls and the use of double glazing to conserve heat.

Overwatering was the biggest problem. Water requirements varied with the plant, and when to water could be determined by feeling the soil or by checking the soil color or the weight of the container. Common sense was to prevail at all times. The soil should never dry out nor the plants wilt before additional water was applied. Sufficient water should be given to thoroughly wet the soil, with the excess accumulating in the saucer being discarded. Tepid rainwater was recommended; ice-cold water caused injury.

Dry air was considered fatal to plants. Central heating raised the temperature and reduced the humidity in rooms. Promotion of atmospheric moisture was essential and could be fostered by syringing the plants, placing pans of water on radiators, or placing the pots over trays of moist sand or cinders.

A well-drained soil was essential. Few natural soils were good; most had to be improved with amendments. A mixture of 2 leaf mold: 1 manure: ½ loam: ½ peat: 1 sand by volume was one recommendation. In potting, 1 to 2 in. of drainage material was suggested for large pots. Soil was to be pressed firmly around roots. Weak fertilizer solutions of soot, ammonia water, or manure could be used sparingly. Liquid manure applied weekly at the rate of 1 tsp per quart of water was recommended.

The importance of ventilation was stressed. Not only were O_2 and CO_2 important, but fumes from gas lights which may have been harmful were dissipated with air exchange.

Victorian homes were dirty, and frequent washing of the foliage was essential. Covering the plants at sweeping time was recommended. Grooming to remove dead leaves and flowers was a routine practice.

Among pests, aphids and red spider were the worst, with mealybugs and scale troublesome at times. Prevention rather than cure was stressed. Washing the plants was the usual control. Others included dusting with red pepper and raising the humidity for red spiders, picking scales, and fumigating aphids with nicotine (tobacco). Control of aphids using nicotine is still an effective control. Dabbing the pest with a feather dipped in alcohol or whiskey would also eradicate them. Use of parasitic wasps to control scale was suggested by Nicholson.

Other problems were usually related to a soil that was too wet or too acid. Leaf drop was associated with drafts, rapid temperature change, overfertilization, and too much or too little water. Once the cause was removed, the effect ceased. Leggy plants were associated with high temperature and low light.

The invention of the electric light provided better interior illumination and helped to stimulate the use of plants indoors, but with the advent of central heating and higher temperatures and lower humidity, plants became difficult to maintain. Popularity declined by the early twentieth century. Palms continued to be popular indoors, but not other plants. Cut flowers became available at low prices, further contributing to the waning interest in indoor plants.

Twentieth Century. As the twentieth century began, foliage plants were hard to get, and very few were used for interior decoration. Cut flowers were less expensive and popular.

In the early years, the Philadelphia area was the foliage capital of the United States. At that time, both Florida and California were relatively undeveloped. Among the plant collectors and producers at that time were W. A. Manda, Robert Craig, and James Vosters. Julius Roehrs, a New Jersey grower of exotic decorative plants, was a major source of plants for greenhouses on private estates of the wealthy. Sylvan Hahn of Pittsburgh was the first to produce *Schefflera actinophylla* as a potted foliage plant, and to grow numerous *Dracaena* species as upright canes.

In the late 1930s, the interest in foliage plants was revived with the introduction of dish gardens. Until the late 1960s, house plants had been used in modest numbers in homes and in public and commercial areas. A change in architecture at the time saw the incorporation of large atriums into public and commercial buildings. The first modern interior plantscape which used plants on a grand scale was installed in 1967 in the Ford Foundation Building in New York. Its success and those that followed can be attributed to Everett Conklin. Called the Dean of Interiorscaping by many, Conklin had the foresight and ability to create many original installations throughout the country.

Thirty years ago, there was virtually no documented experience defining the environmental parameters for long-term maintenance of plants indoors. Such is not the case today. Technological advances, such as improved heating/ventilating/cooling systems and efficient lighting, permit the establishment and maintenance of controlled environments. Coupled with new plant species and innovative cultural techniques, plants can thrive in environments designed for human comfort.

The use of foliage plants has increased tremendously, as interest in and the need for plants indoors have become apparent. The days of the house plant are gone. No longer do we keep a few plants just to watch them grow. Today, plants are used indoors to organize and decorate space. Ornamental woody and herbaceous trees and shrubs of tropical and subtropical origin, adapted for survival in our building interiors, are used to create attractive plantscapes (Figure 2.6).

ORIGINS

Foliage plants presently available for interior planting have been introduced primarily as a result of plant exploration. A few are mutants and others are the result of plant breeding.

Figure 2.6   A hotel lobby.

Exploration

Most foliage plants are tropical or subtropical in origin, and are indigenous to either the tropical belt 1,600 miles north and south of the equator, or the subtropical zone 300 to 600 miles north and south of the tropical zone. These regions have a diversity of climate, not all hot, depending upon elevation, ocean currents, and precipitation, and provide a wide range of plant types suited to indoor culture. Early plant explorations brought many desirable plants to Europe and subsequently to America. Botanists and plant explorers from America have collected and introduced many plants in the past 200 years. The period 1929–1974, however, was the first period of exploration for ornamental plants of economic importance in themselves. Since 1974, recognition of the importance of ornamental plants as a legitimate field for government plant introductions has prompted expeditions to Nepal, the Soviet Union, Japan, Australia, Brazil, India, South Korea, and Europe in search of new plants for indoor and outdoor use. Prior to these missions, many plants were introduced by individuals and the U.S. Department of Agriculture, but they were a sideline to the major intent of the expedition.

Tropical America.   Tropical America has been the source of many of the best plants for interior plantscaping. Some 112,000 species, 60,000 in the Amazon

Basin alone, are estimated to grow in the area. Some frequently used genera include *Aphelandra*, *Beaucarnea*, *Begonia*, bromeliads, cactus, *Calathea*, *Chamaedorea*, *Cissus*, *Columnea*, *Dieffenbachia*, *Episcia*, ferns, *Fittonia*, *Maranta*, *Nephrolepis*, orchids, *Peperomia*, *Philodendron*, *Pilea*, *Spathiphyllum*, *Syngonium*, *Tradescantia*, and *Zebrina*.

Tropical Africa.   More than 40,000 species of plants are indigenous to tropical Africa. For interior plantscaping, the area has provided *Aloe*, *Asparagus*, *Chlorophytum*, *Chrysalidocarpus*, *Coffea*, *Crassula*, *Dracaena*, *Euphorbia*, *Ficus lyrata*, *Kalanchoe*, *Nephrolepis*, *Saintpaulia*, and *Sansevieria*.

Tropical Asia.   The monsoon area of Southeast Asia, including the Philippines, the Himalayas, and Indonesia, is the native habitat of over 50,000 species. Among plants used indoors that came from tropical Asia are *Aglaonema*, *Begonia*, *Coleus*, croton, *Epipremnum*, *Ficus elastica*, *F. benjamina*, *Hoya*, *Nephrolepis*, *Phoenix* and other palms, and certain ferns.

Subtropical Asia.   China and Japan are the source of *Aspidistra*, *Aucuba*, *Citrus*, *Dracaena*, *Euonymus*, *Fatsia*, *Ligustrum*, *Pittosporum*, *Podocarpus*, and *Rhapis*.

Australia and the South Pacific.   *Araucaria*, *Brassaia*, *Cordyline*, cycads, *Dizygotheca*, *Howea*, other palms, and *Polyscias* are among the plants native to this area.

Europe.   Few plants used indoors are indigenous to Europe because of the climate. *Chamaerops humilis*, *Hedera helix*, *Nerium*, and *Solierolia* are the most significant.

North America.   The continent of North America is a minor source of plants for interior plantscaping. *Tolmiea*, various cacti, *Agave*, and *Yucca* are among the native plants used indoors.

Mutants

Some of the foliage plants recommended for interior use are mutants or "sports" of another species. *Dracaena marginata tricolor* is a sport of the green *D. marginata*, while *D. deremensis* 'Warneckii' and *D. deremensis* 'Janet Craig' are mutants of *D. deremensis*. *Dieffenbachia maculata* 'Rudolph Roehrs' arose from *D. maculata*. Many of the cultivars of Boston fern, such as 'Fluffy Ruffles' and 'Fluffy Duffy', are mutants of *Nephrolepis exaltata*. *Dieffenbachia maculata* 'Marianne' and *D. amoena* 'Tropic Snow' are mutants of *D. maculata* 'Perfection' and *D. amoena*, respectively.

Breeding

Jungle exploration and mutants have been the primary source of new plants in the past. Some plant breeding has been attempted, especially with the family *Araceae*, the aroids, which includes *Aglaonema*, *Anthurium*, *Caladium*, *Dieffenbachia*, *Philodendron*, and *Spathiphyllum*. Hybrids of *Schefflera*, Thanksgiving cactus, and ferns are commercially available. *Philodendron* hybrids include 'Emerald Queen' (Figure 2.7), 'Red Princess', 'Emerald King', 'Royal Queen', 'Emerald Duke', 'Black Cardinal',

Figure 2.7   Hybrid *Philodendron* 'Emerald Queen'.

'Prince Dubonnet', 'Majesty', 'Prince of Orange', and others. The very popular *Aglaonema*, 'Silver Queen', is a hybrid, as are 'Fransher', 'Silver King', and 'Parrot Jungle'. 'Sensation' and 'Southern Blush' are *Spathiphyllum* hybrids.

Plant breeding is very expensive and time consuming. One must make a long-term commitment and recognize that success is not guaranteed. As the demand for new plants increases, hybridization, and other genetic manipulation, will become increasingly significant. Breeding of foliage plants is a wide-open field.

Plant breeders will be looking for new or improved plants. Foliage characteristics, plant shape and form, growth habit, resistance to pests and diseases, and tolerance of indoor conditions are among the traits hybridizers will be looking for as their work proceeds. Research is also needed to find efficient ways of producing seed from desirable crosses.

SUMMARY

The history of using container plants indoors is difficult to ascertain. Plants have been used in homes as early as 3,000 years ago, but their extensive use is recent. True pot gardening may have originated with the Greeks in the fifth century B.C., and the custom was copied by the Romans. During the Dark Ages, gardening was preserved by the religious, and it was not until the eleventh and twelfth centuries that an appreciation for decorative plants was revived. The Renaissance was a period of pot gardening. Orangeries, greenhouses, and conservatories were constructed to house new plants introduced as the result of foreign exploration. Exotic plants were collected by the wealthy, a practice that continued until the nineteenth century.

Early in the nineteenth century, the collection of exotic plants became a mark of elegance. Fern cases fostered the mania for ferns. During the Victorian period, the value of plants indoors was recognized, and they were used in profusion.

With the advent of central heating, the popularity declined, to be rekindled in the late 1930s with the introduction of dish gardens. Until the 1960s, house plants were used in modest numbers, but in the last decade the importance of interior plantscaping has become apparent, and their use is here to stay.

Most foliage plants are tropical and subtropical in origin, and have been introduced as a result of plant exploration. A few are mutants. The future lies in plant breeding to provide new and improved plants to satisfy the demand.

REFERENCES

ALLEN, D. E.: *The Victorian Fern Craze*, Hutchinson & Co. (Publishers) Ltd., London, 1969.

BERRALL, J. S.: *The Garden*, The Viking Press, Inc., New York, 1966.

BROWN, C. P.: "Real Native Florida Grown Foliage," *Florida Foliage* 10(12): 12–13, 1984.

CHILDS, E.: "Indoor Gardening," *Nineteenth Century* 4(2): 63–66, 1978.

CONKLIN, E.: "Indoor Landscaping," *Horticulture* 50(3): 48–52, 1972.

"Culture of Marantas," *The Magazine of Horticulture* 31: 81–84, 1865.

CUSHING, J.: *The Exotic Gardener*, W. Bulmer and Co., London, 1814.

DAVIS, W. S.: *A Day in Old Rome*, Allyn and Bacon, Inc., Boston, 1960.

"Dieffenbachia Offer Natural Differences in Size, Appearance," *Greenhouse Manager* 4(12): 25, 1986.

ERMAN, A.: *Life in Ancient Egypt*, Dover Publications, Inc., New York, 1971.

GERARD, J.: *The Herbal or General History of Plants* (1633), reprinted by Dover Publications, Inc., New York, 1975.

GRAF, A. B.: *Exotic Plant Manual*, 2nd ed., Roehrs Company, East Rutherford, N.J., 1970.

GRIFFITH, L.: "The New Hybrid Philodendron," *Florists' Review* 173(4481): 54–55, 1983.

HAMMER, N.: "The Roots of Interiorscaping," *Interiorscape* 12(4): 22–25, 1993.

"Handsome Foliage Plants," *Horticulturist* 30: 137, 1875.

HENDERSON, P.: *Practical Floriculture*, Orange Judd and Company, New York, 1869.

HENNY, R. J.: "Foliage Plant Breeding—Past, Present, Future," *Proceedings of the 1977 National Tropical Foliage Short Course*, pp. 173–175.

____, and E. M. RASMUSSEN: *"Aglaonema* Hybridization Guide," *Nurserymen's Digest* 18(6): 74–75, 1984.

HIBBERD, S.: *The Amateur's Greenhouse and Conservatory*, Groombridge and Sons, London, 1883.

____, *New and Rare Beautiful Leaved Plants*, Bell and Daldy, London, 1870.

____, *Rustic Adornments for Homes of Taste*, Groombridge and Sons, London, 1870.

HUXLEY, A.: *An Illustrated History of Gardening*, Paddington Press Ltd., New York and London, 1978.

HYAMS, E.: *A History of Gardens and Gardening*, Praeger Publishers, Inc., New York, 1971.

"Interior Landscaping," *Florists' Review* 140(3632): 23ff., 1967.

JASHEMSKI, W. F.: *The Gardens of Pompeii*, Caratzas Brothers Publishers, New Rochelle, N.Y., 1979.

JOURDAIN, M.: *English Interior Decoration (1500–1830)*, B. T. Batsford Ltd., London, 1950.

KETHRELL, R. H., ed.: *Early American Rooms (1650–1858)*, Dover Publications, Inc., New York, 1967.

KILBORN, J.: "A Victorian Heritage," *Horticulture* 59(1): 49–50, 52–55, 1981.

KRAFT, K., and P. KRAFT: "House Plants," *Americana* 4(6): 46–49, 1977.

LAMANCE, L. S.: *House Plants*, John Lewis Childs, Floral Park, N.Y., 1893.

LAWRENCE, G. H. M.: "Horticulture," in *A Short History of Botany in the United States*, Joseph Ewan, ed., Hafner Publishing Company, Inc., New York, 1969, pp. 132–145.

MALING, E. A.: *In-door Plants and How to Grow Them*, Smith, Elder and Co., London, 1862.

MCINTOSH, C.: *The Greenhouse, Hot House and Stove*, Wm. S. Orr and Co., London, 1840.

MENEN, A.: "The Rise and Fall of the Roman House Plant," *Horticulture* 55(12): 27–29, 1977.

MOLLISON, J. R.: *The New Practical Window Gardener*, Groombridge and Sons, London, 1879.

MURRAY, R.: "On the Management of Plants in Rooms," *The Magazine of Horticulture* 2: 11–13, 1836.

NICHOLSON, G., ed.: *The Illustrated Dictionary of Gardening*, L. U. Gill, London, 1887.

"On Growing Plants in Rooms," *The Magazine of Horticulture* 27: 133, 1861.

"Palms as Decorative Plants," *The Magazine of Horticulture* 34: 273–277, 1868.

"Parlour Culture—Ferns," *Horticulturist* 14: 9–11, 1859.

PETERSON, H. L.: *Americans at Home*, Charles Scribner's Sons, New York, 1971.

"Plants for Parlour and Conservatory," *Horticulturist* 28: 92, 1873.

RAND, E. S., Jr.: *The Window Gardener*, Henry L. Shepard and Co., Boston, 1874.

____: *Flowers for the Parlor and Garden*, J. E. Tilton and Co., Boston, 1864.

RHEE, K.: "The Ascent of Houseplants," *Horticulture* 57(1): 62–67, 1979.

RINGLER, B.: "A Bit of History," *House Plant Magazine* 2(2): 47, 1993.

ROBINSON, W.: *The Parks, Promenades and Gardens of Paris*, John Murray, London, 1869, pp. 262–311.

ROEHRS, J.: "How It All Began," *Interiorscape* 8(4): 66, 52, 1989.

ROGERS, M. R.: *American Interior Design*, W. W. Norton & Company, Inc., New York, 1947.

RYERSON, K. A.: "Plant Introductions," *Agricultural History* 50(1): 248–257, 1976.

SMITH, C. N., and E. F. SCARBOROUGH: "Status and Development of Foliage Plant Industries," in *Foliage Plant Production*, J. N. Joiner, ed., Prentice-Hall, Inc., Englewood Cliffs, N.J., 1981, pp. 1–39.

SWEET, R.: *The Hot-House and Greenhouse Manual*, James Ridgway, London, 1831.

WEISENBERG, L. A.: "Terrariums, Greenery under Glass," *Americana* 2(5): 21–23, 1974.

WHITE, M. G.: *Pots and Pot Gardens*, Abelard-Schuman Ltd., London, 1969.

WILFRET, G. J., and T. J. SHEEHAN: "Development of New Foliage Plant Cultivars," in *Foliage Plant Production*, J. N. Joiner, ed., Prentice-Hall, Inc., Englewood Cliffs, N.J., 1981, pp. 126–136.

WILLIAMS, H. T., ed.: *Window Gardening*, 12th ed., Henry T. Williams, New York, 1876.

"Within Doors," *Horticulturist* 21: 321–324, 1866.

WRIGHT, R.: *The Story of Gardening*, Dodd, Mead & Company, New York, 1934.

Light

Light is a major determinant of plant growth and development, influencing many plant processes.

THE ROLE OF LIGHT

Photosynthesis.   Photosynthesis, the production of food (sugar) from carbon dioxide and water in the presence of chlorophyll and light, is probably the most important of all photochemical processes. Occurring only in certain cells of green plants, photosynthesis generates the food on which all life depends. In addition, oxygen is produced in the process.

Chlorophyll Synthesis.   Chlorophyll, the green pigment of plants and absorber of light for photosynthesis, requires light for its manufacture. In the absence of light, leaves will not be green. High light intensities destroy chlorophyll in some plants.

Stomatal Behavior.   Stomates, pores in the leaf that permit gas exchange, are influenced by light, and are typically open during light periods and closed in the dark, except in succulents with crassulacean acid metabolism, where stomates are open at night and closed during the day. In either case, when the stomates are open, carbon dioxide and oxygen are able to diffuse rapidly into and out of the leaf. Water vapor lost in the process of transpiration also passes out of the plant, predominantly through the stomates in the leaf surface.

Photomorphogenesis.   Light may also have a profound influence on the development of various plant parts. Termed photomorphogenesis, light may influence flower initiation and development, seed germination, and the onset of the dormant or resting period. Photomorphogenic responses occur mainly in red and far-red light, and may be associated with photoperiod, or length of the day.

Anthocyanin Production.   Anthocyanin, a red pigment, depends upon light for the production of sugar, one of its building blocks. In bright light, more photosynthate is produced and red-leaved plants such as croton, *Coleus,* and *Zebrina* will be more spectacular.

Increase in Temperature.   Light usually warms plant surfaces and thereby increases the rate of most physiological processes, including photosynthesis, respiration, and transpiration. Excessively high temperatures may cause injury.

Translocation.   Translocation, the movement of materials throughout the plant, is influenced by light. In light, food is translocated mainly to the stem tips and other shoot parts; in the dark, translocation to the root system predominates.

Mineral Absorption.   Mineral absorption by roots is faster in high light, probably because more photosynthate is available to provide the energy needed for active ion uptake.

Abscission.   Abscission, or dropping, of plant parts, especially leaves, is influenced by light intensity. A *Ficus benjamina* plant produced in high light may lose more than half of its leaves if placed directly in the low light of most building interiors, for example, because reduced photosynthesis starves the tissue.

Plants respond to three aspects of light, the intensity or brightness, the duration or day length, and the quality or color.

LIGHT INTENSITY

Intensity refers to the brightness, or level of light. In most instances, it is the limiting factor in growing and maintaining plants inside.

The light intensity indoors will vary depending on the source of the light, obstructions, and reflection. Natural light, the sun, may provide most, some, or none of the light for a building interior. The intensity of the daylight varies with the season of the year and with distance north or south of the equator. It is brightest in summer and nearer the equator. Cloud cover and composition; atmospheric dust, moisture, and haze; the elevation on the earth; and the plane of exposure also influence the brightness of sunlight at the earth's surface. The amount of natural light entering a building will be influenced by the area and exposure of the glass surface, the presence of trees and shrubs, roof overhangs, window screens and awnings, and the color and cleanliness of the glass. A $1/4$-in.-thick pane of bronze glass allows 21% visual light transmission; gray, 41%; and clear, 89%.

Inside, the amount of natural and/or artificial light available for the plants will be affected by curtains and blinds, surface textures, and reflectance from wall coverings, draperies, and furniture. White plaster reflects about 90%; mirrors, 80 to

90%; gray-beige paint, 50%; and draperies, 35%. When properly designed, an artificial lighting system would provide the most uniform light intensity.

Measurement

Determination of the light intensity on the area where plants are to be used should not be guesswork. An accurate light meter should be used to measure the light level. If natural light or a combination of natural and artificial light is used, several readings should be taken at various times of the day and at different seasons. An average of these values will be the light intensity for the area.

Light intensity may be expressed in any of several units: the footcandle and the lux are commonly used. A footcandle (fc) is the unit of illumination equivalent to that produced by a standard candle at a distance of 1 ft. Footcandles will be used in this text to designate light intensity. Lux is the international unit of illumination and is the illumination received by a surface at a distance of 1 meter (m) from a light source whose intensity is taken as unity. One lux is equal to 0.09 fc.

The simplest way to measure light intensity is with a direct-reading light meter. The Model 214 manufactured by General Electric (Figure 3.1) is a pocket-size meter which permits one to obtain light readings up to 10,000 fc. The cost is about $75. Although this meter is most sensitive to the green-yellow light to which the human eye responds, it is accurate enough for assessing the light used by plants. Photometers which are more accurate are extremely expensive.

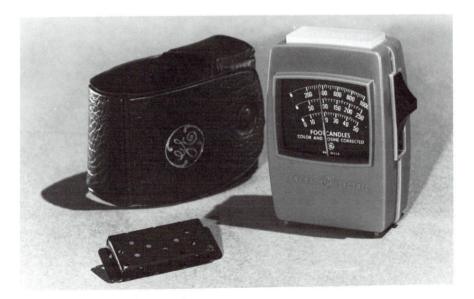

Figure 3.1   Direct-reading, pocket-size light meter. (Courtesy of General Electric Company.)

Avoid exposing the light meter to bright light as damage to the photocell may occur. Start with the meter set for the brightest light and work down until the needle moves onto a scale that can be read.

Photographic light meters may also be used by converting the light reading to footcandles. One method of conversion is as follows:

1. Set the meter as ASA 100.
2. Aim the photocell at a white sheet of paper held parallel to the plane of the leaf.
3. Hold the photocell no farther away from the paper than its shortest dimension. (For $8\frac{1}{2}$- by 11-in. paper, the light received should be $8\frac{1}{2}$ in. away.)
4. Set the f-stop at 4.
5. The shutter speed is the approximate light intensity in footcandles.

The foregoing procedure will measure the visible spectrum, and is accurate to within 20%. Other conversions using photographic light meters are available. The owner's manual for the particular meter or camera with built-in meter may contain a conversion table. If not, the manufacturer can provide one.

Without a light meter, accurate assessment of the light intensity is not possible. There are a few guidelines, however, which may be helpful in making a value judgment when a light meter is not immediately available. For example, if you can see the shadow of your fingers when your hand is held 12 in. above a piece of white paper, the light intensity is at least 100 fc, certainly sufficient for the maintenance of most foliage plants. Table 3.1 shows the recommended average light intensities for ordinary locations. Examples of the illumination produced from various lamps used in plant lighting is presented in Table 3.2. Light intensity falls off rapidly as the distance between the lamp and the plant increases, and is highest under the lamp.

Plant Response

There is a minimum light intensity for plant maintenance, which varies with each species. The level of light required is related to the native habitat of the plant, the light intensity at which it was produced, and the degree of acclimatization. It is less than that necessary to promote growth (Table 3.3). Plants indigenous to tropical rain forests will have a lower requirement than those which normally grow in open woodlands or full sun. Plants with colored and variegated leaves require higher light intensity than green-leaved forms. Flowering plants require brighter light than those used for their foliage alone. In this book, we are concerned primarily with plant maintenance, not growth. Plants are to produce new leaves at a rate sufficient to replace leaves lost in the normal aging process. They are not to be grown from small plants to large specimens.

The minimum light intensity for maintenance of a foliage plant is that which permits the plant to function at a level slightly exceeding the compensation point at which photosynthesis (food manufacture) is equal to respiration (food utilization).

TABLE 3.1   Average Light Intensities for Ordinary Locations

Location	Footcandles of Light
Outdoors	
Sunny day (summer)	To 10,000
Cloudy winter day	500–2000
Offices	
Conference rooms	20–30
Typing and accounting	40–50
Drafting room	50–100
Drafting-table top	100–200
Reception desk	10–30
Homes	
Living room	
Day	10–1000
Night	5–10
Reading	20–30
Workbench	40–60
Sewing	50–100
Stores	
Circulation area	20
Merchandise area	30–100
Displays	100–300
Churches	
Foyers	2–15
Chapels	5–10
Halls	< 5

Source: "Landcape Design Goes Indoors with Plants," Grounds Maintenance 2(4): 29–31, 1967.

Below the compensation point, the plant will draw on its food reserves and deteriorate very quickly. At the compensation point, the plants will die in less than 2 years, depending on the species. New leaves are not produced, but old leaves have a life span and will senesce and abscise. With fewer and fewer leaves, the attractiveness of the plant will decline, and ultimately it will die. *Ficus benjamina* may last 2 years in such an environment, and *Dieffenbachia* must be replaced within 1 year.

To maintain a high-quality interior plantscape, lost foliage must be replaced. At light intensities above the minimum (Table 3.4), new leaves will be produced. Intensities higher than the preference levels will cause the plants to grow, and pruning will be required.

The light-compensation points for most foliage plants are not known—hence the conflicting information one frequently finds when consulting several sources for the light requirement of a specific plant. Fonteno and McWilliams (1978), in Texas, studied the light-compensation points of four tropical foliage plants: *Philodendron scandens* 'oxycardium', *Epipremnum aureum*, *Brassaia actinophylla*, and *Dra-*

TABLE 3.2 Illumination (Footcandles) at Various Distances from Several Lamps[a]

Distance from Lamp (ft)	Incandescent				Fluorescent			
	Standard (Double Values With Reflectors)		Par 38		40-W U-bulb With Reflector		Standard 40-W	
	40-W	60-W	75-W	150-W	One Bulb	Two Bulbs	Channel	Reflector
1	34 (17)	67 (33)	—	—	240	300–400	260 (200)	400 (260)
2	8 (7)	17 (13)	—	—	80	100–200	100 (100)	180 (150)
3	4 (3)	7 (7)	375 (40)	383 (80)	40	50–100	60 (60)	100 (90)
4	—	—	167 (40)	216 (110)	—	—	40	60
5	—	—	94 (50)	138 (90)	—	—	—	—
6	—	—	60 (40)	96 (70)	—	—	—	—

[a]Values in parentheses are the footcandles 1 ft on either side of lamp, perpendicular to distance below lamps.

Source: Adapted from H. M. Cathey and L. E. Campbell, *Indoor Gardening*, 1978, p. 14.

TABLE 3.3   Recommended Light Intensities for Maintenance and Growth of
Selected Foliage Plants (fc of Cool-White Fluorescent Light, 16 hrs/day)

Species	Maintenance	Growth
Asparagus setaceus	100–150	200–400
Brassaia actinophylla	150-200	200-400
Chamaedorea elegans	50–75	75–100
Chlorophytum comosum	100-200	200–400
Cissus antarctica 'Minima'	100–150	150–400
Cissus rhombifolia	100–150	150–400
Cordyline terminalis	75–100	100–150
Dizygotheca elegantissima	50–100	100–200+
Dracaena marginata	100–150	200+
Epipremnum aureum	75–100	100–200+
Ficus benjamina	100–150	150–400+
Hedera helix 'California'	100–150	150–300
Nephrolepis exaltata 'Bostoniensis'	50–100	150–200+
Philodendron scandens 'oxycardium'	50–100	100–200+
Podocarpus macrophyllus	100–200	400+
Syngonium podophyllum	50–100	100–150

Source: Adapted from A. M. Kofranek, Proceedings of Environmental Conditioning Symposium, 1977, p. 8.

TABLE 3.4   Suggested Light Intensities for
Maintenance of Foliage Plants Indoors

Requirement of Particular Species	Minimum (fc)	Preferred (fc)
In sun		
Low	12	35–100
Medium	35	100–250
High	100	250
Very high	500	> 500
Artificial		
Low	25	75–100
Medium	75–100	200–500
High	200	500
Very high	1000	> 1000

Source: H. M. Cathey and L. E. Campbell, Indoor Gardening,
1978, p. 18.

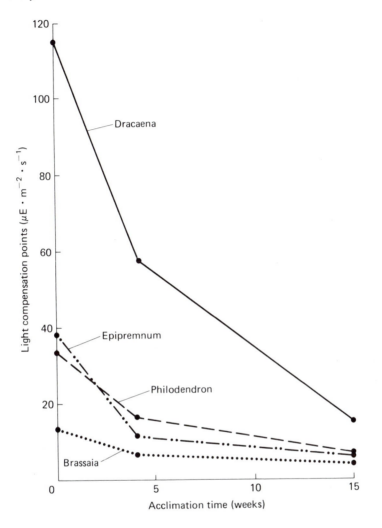

Figure 3.2 Relative light-compensation-point response of four foliage plant species during acclimatization. [From W. C. Fonteno and E. L. McWilliams: "Light Compensation and Acclimatization of Four Tropical Foliage Plants," *Journal of the American Society for Horticultural Science* 103(1): 52–56, 1978.]

caena sanderana. Their data showed a decrease in light-compensation points with increased acclimatization time (Figure 3.2), accompanied by an increase in net CO_2 uptake and 50 to 70% decrease in dark respiration. They would not suggest a minimum light intensity for each of the four species, and stated that factors not studied may have contributed to a lower light-compensation point. In terms of intensity, the light-compensation points determined for the respective plants were 39, 33, 22, and 83 fc. Earlier, Bohning and Burnside (1956), using eight sun species, including *Coleus*, and five shade species, including *Nephrolepis exaltata* 'Bostoniensis', *Saintpaulia ionantha*, and *Philodendron cordatum* (*scandens* 'oxycardium'), showed that light saturation and compensation points occurred at higher intensities in sun than in shade plants, and were 100 to 150 fc and 50 fc, respectively.

Similar results have been reported by others. Pass and Hartley (1979) found increased rates of net CO_2 uptake and decreased rates of dark respiration at low irradiances in *Brassaia actinophylla*, *Nephrolepis exaltata* 'Bostoniensis', and *Epipremnum aureum*. *B. actinophylla* had the greatest photosynthetic efficiency at low light intensities, followed by *E. aureum* and *N. exaltata* 'Bostoniensis'. Fails et al. (1982) and Nell et al. (1981) reported that shade-grown *Ficus benjamina* had lower light-compensation points with higher net photosynthesis and lower dark respiration indoors than sungrown plants. Braswell et al. (1982) showed that *Schefflera arboricola* maintained better quality than *B. actinophylla* when held under low light levels.

Generally speaking, if a plant is maintaining leaves, the light level is adequate, and if a new leaf is formed, the oldest leaf stays on the plant. Leaves are parallel to the light source, have the proper size and density, and are dark green with a thick and crinkly texture. The brighter the light, the shorter the internodes. If the light intensity is too low, leaves may yellow and drop from the plant (Figures 3.3–3.5). New growth will be leggy, with weak, elongated stems and wide internodes (space between leaves). New leaves will be small. If the light is too bright, leaves may scorch. New leaves may be small, bleached, and curled. Scorched,

Figure 3.3 *Ficus benjamina* moved directly indoors from a full-sun production area: (left) first day; (right) 6 weeks later.

Figure 3.4 *Aglaonema commutatum* 'Silver Queen'. Too-low light reduces variegation and causes paling of leaves (left); proper light (center); high light will bleach foliage and may cause burning (right).

bleached leaves will not return to normal if the plant is moved to a more favorable light level (Figure 3.6).

Within physiological limits, increasing the light intensity will increase photosynthesis and reduce the number of hours of light the plant must receive each day. If the plant receives light from one side, it will bend toward that side. Turning the plant periodically will facilitate straight growth.

Selecting Plants

In selecting plants for an interior landscape, two approaches are possible. Either select plants for the given light intensity, or change the light intensity to suit the plants one has chosen. The public is often misled by seeing healthy-looking plants in areas where there is insufficient light. At light levels below 30 fc, live plants are not recommended because they deteriorate very rapidly. They may be used in these locations only if replaced frequently with new plants. Use of preserved or artificial plants may also be justified in low-light areas. Table 3.5 lists the longevity of certain foliage plants at various light intensities. Kofranek (1977) studied the maintenance of 12 species under three intensities of fluorescent light for 4 months. *Codiaeum* and glacier ivy did very poorly even under high light (100 to 175 fc). All others, *Aechmea fasciata*, *Aglaonema commutatum*, *Aphelandra squarrosa* 'Louisae', *Chamaedorea elegans*, *Dieffenbachia maculata* 'Rudolph Roehrs', *Ficus lyrata*, *Maranta leuconeura*

(a)

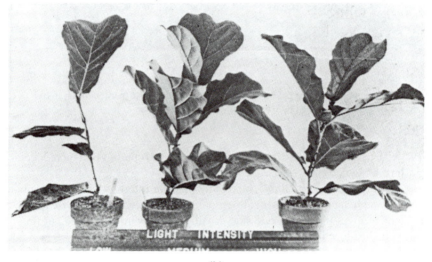

(b)

Figure 3.5   (a) *Aechmea fasciata* and (b) *Ficus lyrata* after 4 months at low (25 to 30 fc), medium (45 to 90 fc), and high (100 to 175 fc) light. [From A. M. Kofranek: "The Maintenance of Some Indoor Foliage Plants under Fluorescent Lighting," *Florists' Review* 150(3895): 19–20, 1972.]

'Kerchoveana', *Peperomia obtusifolia, Philodendron pertusum (Monstera deliciosa)*, and *Brassaia actinophylla* did well under high light, and could have been maintained satisfactorily for more than the 4 months of the tests.

Information about the light requirements for specific plants may be found in numerous books which describe the culture of the myriad plants used indoors. In Chapter 12, a list of plants recommended for interior landscaping is presented together with the suggested light intensities.

Figure 3.6   Too-bright light has destroyed the chlorophyll and bleached this *Fuchsia* leaf. The tissue may later die, resulting in a scorched appearance.

LIGHT SOURCES

The Sun

The sun is the universal source of energy for the earth. It is an incandescent source, with light produced by thermal radiation. About one-half of the sun's energy reaches the earth's surface. Most of the infrared (heat) irradiation is absorbed by the carbon dioxide (CO_2) and most of the ultraviolet irradiation is absorbed by the ozone and oxygen in the upper atmosphere. Ultraviolet light degrades and denatures the proteins and nucleic acids in living cells and, if it were not absorbed, there would be no life.

Where practical, as much natural light as possible should be used for the indoor garden, since doing so saves energy. As discussed previously, daylight is extremely variable, and both natural and physical factors must be considered in assessing the light level available from the sun.

Natural light falls off very rapidly with distance from the window, with usable light generally not penetrating much beyond 15 ft from the glass (Figure 3.7). To maximize the utilization of light from sidewall windows, plants should be placed within a 45° area of the window head (Figure 3.8).

If skylights are to be installed to provide natural light, they should be properly designed, with the well as wide and shallow as possible. Beveling the wall of the well and painting it flat white will increase lighting efficiency. On sunny days, light

TABLE 3.5   Number of Months Foliage Plants Will Remain Attractive Under Various Light Intensities

Foliage Plant	Light Intensity in Footcandles (16 hrs/day)			
	15–25	25–50	50–75	75–100
Aglaonema commutatum	12	36	36	
A. modestum				36
Araucaria heterophylla		36	38	38
Aspidistra elatior	12			
Aucuba japonica	12		36	38
Brassaia actinophylla		30	36	38
Chlorophytum spp.		30		36
Cissus rhombifolia				12
Dieffenbachia amoena	12		36	
D. maculata cv. Baraquiniana	12	12	12	
D. maculata 'Rudolph Roehrs'		12		
Dracaena deremensis 'Warneckii'	30	36	36	38
D. fragrans massangeana			30	36
D. sanderana	12			
Epipremnum aureum			30	36
Ficus benjamina 'Exotica'				12
F. elastica 'Doescheri'				12
F. lyrata				12
Hedera helix 'Marble Queen'				12
Howea forsterana		12		
Nephrolepis exaltata 'Bostoniensis'		12		
Peperomia obtusifolia		12		
Philodendron bipennifolium				34
P. domesticum	12	30	36	36
P. pertusum (Monstera deliciosa)				36
P. scandens subsp. oxycardium	12	24		
Pilea cadierei		12		
Polyscias balfouriana marginata			12	
Sansevieria spp.	12			
Spathiphyllum cv. Clevelandii			12	
Syngonium podophyllum	12			
Tolmiea menziesii				12

Source: Adapted from G. Thames and M. R. Harrison, Foliage Plants for Interiors, Bulletin 327-A, Rutgers University, New Brunswick, N.J., 1966, pp. 22–23.

intensity in areas with frosted (translucent) skylights is four to five times higher than with clear because light rays are scattered by the frosted panels. On cloudy days, there is no difference. In greenhouse-type structures, high light intensity and excessive heat may injure plants by scorching leaves. Shading to reduce the light and providing adequate ventilation will correct the problem.

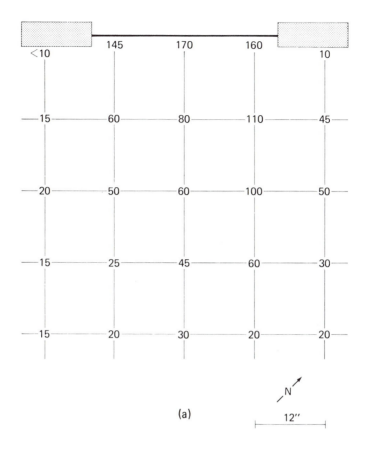

Figure 3.7   Light intensities within a room at various distances from clear, unobstructed windows.

Artificial Light (Lamps)

Artificial light may be used as a supplement to natural light which may not be bright enough, or as the sole source of illumination. As a supplement, the amount required will depend upon the kinds of plants and the level of natural light available. Lamps are burned for fewer hours than when they are the sole source, and may be used at any time of the day. Regardless of whether the light is used as a supplement or as the sole source, the level needed will be determined by the plants with the highest light requirement.

Artificial light may serve a dual role: room lighting and plant lighting. General room lighting may not be adequate for plant maintenance; however, supplementary light in the area where the plants are located should solve any problems. For

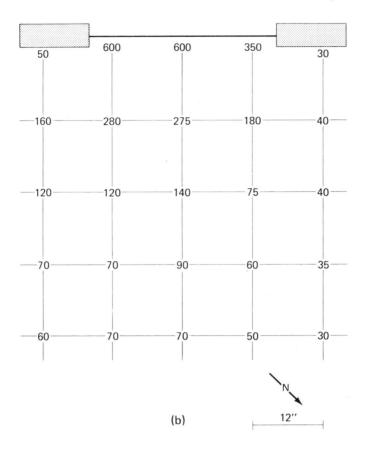

(b)

Figure 3.7   (continued)

example, if the general lighting is 25 fc, and 100 fc is required for a particular plant, add 75 fc of supplemental light over the plant.

It is impractical and unnecessary to try to duplicate the intensity of the sun, as the photoresponses of plants occur at much lower light levels. The light used may or may not simulate the spectral-emission discharge (S.E.D.), or color, of sunlight. For maintenance, however, the color must be satisfactory for both plants and people (Table 3.6). Fluorescent plant lamps, except those which are full-spectrum, are generally unsatisfactory for general lighting. Any "white" source will usually be adequate.

Supplemental light for foliage plants must be efficient in the region of the spectrum from 500 to 800 nanometers (nm). This is primarily the visible spectrum, encompassing blue at the lower end and red and far-red at the upper (Figure 3.9). All artificial light sources can serve as plant lamps. The types include tungsten fila-

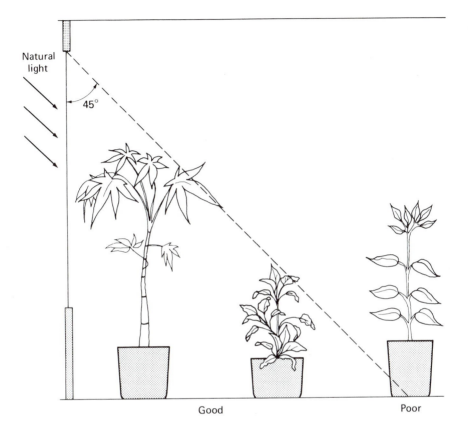

Figure 3.8 Suggested placement of plants to maximize utilization of light from a sidewall window.

ment incandescent lamps, and gas discharge lamps such as fluorescent, mercury, metal halide, and sodium. The lamp or lamps for the indoor garden must be selected with consideration of the specific responses wanted: for example, control of stem elongation, branching, flowering, or foliage color (Table 3.7).

Another consideration in the selection of a lamp is light efficiency. Incandescent lamps radiate less physiologically active light than do discharge lamps such as fluorescent or high-pressure sodium. Sodium lamps require heavy fixtures and large ballasts and burn so brightly that they have limited use in small spaces such as homes or small shops. Where ceiling height is less than 12 ft, fluorescent lamps are probably the best choice.

Incandescent Lamps. Incandescent lamps are the most widely used method of lighting and produce light when a current flows through a tungsten filament heating it to incandescence. They were developed by Thomas A. Edison over 100 years ago,

TABLE 3.6 Color Rendering of Plants, People, and Furnishings Produced by Various Light Sources

Lamp	General Appearance on Natural Surfaces	Complexion	Atmosphere (Feeling in Room)	Colors Improved	Colors Grayed (Undesirable)
Incandescent	Yellowish white	Ruddy	Warm	Yellow, orange, red	Blue
Mercury incandescent	Yellowish white	Ruddy	Warm	Yellow, orange, red	Blue
Fluorescent					
Cool white	White	Pale pink	Neutral to cool	Blue, yellow, orange	Red
Warm white	Yellowish	Sallow	Warm	Yellow, orange	Blue, green, red
Gro-Lux	Pink-white	Reddish	Warm	Blue, red	Green, yellow
Gro-Lux WS	Light pink-white	Pink	Warm	Blue yellow, red	Green
Agro-lite	White	Pink	Warm	Blue, yellow, red	Green
Vita-Lite	White	Pink	Warm	Blue, yellow, red	Green
Mercury	Purplish white	Ruddy	Cool	Blue, green, yellow	Red
Metal halide	Greenish white	Grayed	Cool green	Blue, green, yellow	Red
High-pressure sodium	Yellowish	Yellowish	Warm	Green, yellow, orange	Blue, red
Low-pressure sodium	Yellowish	Grayed	Warm	Yellow	All but yellow

Source: H. M. Cathey and L. E. Campbell, Interior Landscape Industry 1(4): 37, 1984.

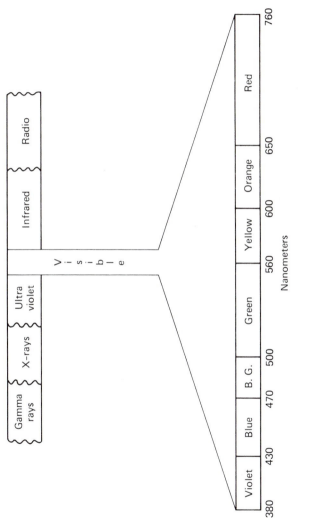

Figure 3.9 Electromagnetic spectrum. The band between ultraviolet and infrared is visible light. (Courtesy of GTE Lighting Products.)

TABLE 3.7   Lamps and Plant Response

Lamp	Plant Response
Fluorescent Cool white (cw) and warm white (ww)	Green foliage expands parallel to the surface of the lamp. Stems elongate slowly. Multiple side shoots develop. Flowering occurs over a long period of time.
Plant lamps—Gro-Lux	Deep green foliage which expands, often larger than on plants grown under cw or ww. Stems elongate very slowly, extra-thick stems develop. Multiple side shoots develop. Flowering occurs late, flower stalks do not elongate.
Wide spectrum—Gro-Lux WS, Vita-Lite, Agro-lite	Light green foliage which tends to ascend toward the lamp. Stems elongate rapidly, large distances between leaves. Suppresses development of multiple side shoots. Flowering occurs soon, flower stalks elongated, plants mature and age rapidly.
High-intensity discharge Deluxe mercury or metal halide	Similar to cw and ww fluorescent compared on equal energy. Green foliage which expands. Stems elongate slowly. Multiple side shoots develop. Flowering occurs over a long period of time.
High-pressure sodium	Similar to Gro-Lux and other color-improved fluorescent compared on equal energy. Deep green foliage which expands, often larger than on plants grown under mercury or metal halide. Stems elongate very slowly, extra-thick stems develop. Multiple side shoots develop. Flowering occurs late, flower stalks do not elongate.
Low-pressure sodium	Extra-deep-green foliage, bigger and thicker than on plants grown under other light sources. Stem elongation slowed, very thick stems develop. Multiple side shoots develop, even on secondary shoots. Flowering occurs, flower stalks do not elongate.
Incandescent and incandescent mercury	Paling of foliage, thinner and longer than on plants grown under other light sources. Stem elongation is excessive, eventually stems become spindly and break easily. Side-shoot development is suppressed, plant expands only in height. Flowering occurs rapidly, the plants mature, and senescence takes place. Exceptions: Rosette and thick-leaved plants such as *Sansevieria* may last several months. New leaves will elongate and will not be typical of the species.

Source: H. M. Cathey and L. E. Campbell, *Indoor Gardening*, 1978, pp. 16–17.

and have remained relatively unchanged since that time. Ordinary household light bulbs are examples of incandescent lamps. Such lamps produce a considerable amount of red light but are low in blue light.

Several types of lamps are available, including Standard (A), Reflector (R), Cool Beam (PAR), and plant growth, either clear or frosted. Bulb shape varies, as shown in Figure 3.10; the screw-base type is the most common.

The household light bulb is an example of a standard (A) incandescent lamp. They are available in a wide range of wattages and colors. Reflector (R) lamps have an inside reflecting surface which cannot accumulate dirt. They are available as spot or flood lamps in various wattages. R-lamps conduct heat upward away from the plants. PS-30 reflectorized lamps eliminate the characteristic hot spot of R-lamps and provide a wide flood distribution. Both standard and R-lamps must be protected from moisture, as the glass may shatter if water hits them when they are burning. PAR bulbs are made of thick glass and are more rugged and weather-resistant than either of the first two. They have a built-in reflector and are available in several wattages. PAR lamps are more efficient than R-lamps and have a more precise beam

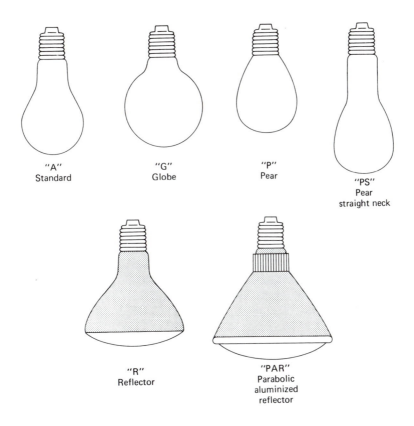

Figure 3.10   Some incandescent lamp types suitable for indoor gardens.

pattern—narrow, flood, or wide-flood. Specially designed incandescent plant lights are available as 75- or 150-W reflector bulbs, and 60- to 100-W nonreflector lamps. They have a blue tint, which allegedly corrects the inherent blue-light deficiency of standard incandescent lamps, thus providing more natural illumination, but reducing light intensity. These incandescent plant lamps are very expensive but are no better than standard incandescent lamps.

Incandescent lighting requires simple circuitry compared with other types. No special equipment is needed. Ceramic sockets are recommended for all installations, and are required when PAR lamps are used, as less durable sockets may cause poor contact and tend to short out after several months. Plastic sockets may melt as a result of the heat produced. All incandescent lamps generate considerable heat, which may injure plants. Placing plants at some distance from the lamp reduces this possibility. Heat shields may be used, but they also reduce light intensity.

There are many advantages to using incandescent light, particularly as supplemental lighting for plants.

1. It is a compact light source, with good control of light output.

2. Initial installation costs are low: no expensive fixtures are required.

3. Simple circuitry requires no ballast.

4. Light output is not a function of ambient temperature.

5. Light output is not a function of burning hours per start.

6. For the size of the lamp, there is a high light output.

7. Reflector flood lamps and spot lamps may be used to create dramatic shadows and illuminate plants. An incandescent beam carries better than does diffuse fluorescent light.

8. The lamps fit into attractive fixtures.

9. Lamps are available in a wide range of wattages.

There are certain disadvantages associated with incandescent lamps which may prohibit their use in most situations.

1. They are not energy-efficient and have low light output per input watt. About 6 to 8% of the energy emitted by an incandescent lamp is physiologically active. Efficiency increases as lamp wattage increases, making it possible to save energy and fixture costs by using higher-wattage lamps. One 100-W standard lamp produces more light than two 60-W standard lamps, for example. The light produced is high in orange and red and low in blue.

2. The lamps produce a lot of heat, 84% of the energy generated.

3. Light output is affected by voltage variations, which alter the temperature of the filament. Voltages above the rated voltage yield more light but shorten lamp life and use more energy. Lower voltages yield less light, extend lamp life, and use less energy. The S.E.D. also changes with voltage differences.

4. Lamps have a relatively short life: 750 hrs. Long-life lamps have a different S.E.D. than standard lamps of equal wattage and produce less light. Their use is not recommended.

5. Lamps act as a point source of light, and distribution on a flat surface under the lamp is not uniform.

6. Intensities are not sufficiently high for flowering plants.

7. There is a loss of efficiency with time. Light output diminishes with time as a result of blackening of the glass and lower filament temperature. Lamps should be replaced after 500 hrs of use.

In certain situations, it may be necessary to adjust the intensity of incandescent lighting. This should be done by changing the wattage of the lamp or the distance between the plant and the lamp. Dimmers are not recommended because they change the S.E.D. of the lamp and interfere with the normal photoresponses of the plant. Neutral density filters may also be used to alter light intensity.

A variation of the incandescent lamp is the tungsten-halogen (quartz) lamp. These lamps use a halogen regenerative cycle which minimizes lamp blackening, thus maintaining excellent light output.

Tungsten-halogen lamps produce white light and are available as tubular, PAR, and R-bulbs in wattages ranging from 75 to 1500 for use in general, spot, and flood lighting. Lamps have twice the life of comparable-wattage incandescent lamps with the same lumen output and are more compact for a given power rating. About 10 to 12% of the light is in the visible spectrum. Lamps are available either single- or double-ended with bases that are compatible with those used in conventional incandescent lamps.

Self-ballasted incandescent-mercury lamps are relatively new. These lamps provide the color balance of fluorescent, and the bright light of mercury vapor but without the need for the ballast. Incandescent-mercury lamps are available as either standard or flood lamps in 160- and 250-W sizes which replace 200- and 300-W incandescent lamps, respectively. They fit into standard ceramic sockets and outlast standard incandescents by more than 12 times. High cost of the lamps is the major drawback.

Incandescent lamps are considerably less efficient than other light sources and generate tremendous amount of heat, which can injure plants. Primarily for these reasons, they have limited use indoors. If no other light source is available, however, they are better than nothing.

Fluorescent Lamps. Fluorescent lamps are low-pressure mercury, electric discharge lamps. The light is produced by the action 253.7-nm radiation from a low-pressure mercury arc on the phosphor that coats the inner surface of the tube. The phosphor converts the short waves to the longer-wavelength radiation of light. The transmission characteristics of the glass and the emission characteristics of the phosphor determine the color. Fluorescent light is high in blue and yellow-green and low in red light. Cool-white fluorescent lamps rend "true" plant color and are the standard by which other lamps are measured.

Fluorescent lamps are available in wattages ranging from 15 to 215, and lengths ranging from 6 to 96 in. The basic lamp consists of a phosphor-coated glass tube into which is sealed an inert gas at low pressure, a small amount of mercury, and a cathode at each end. Externally, a base, usually contact pins, completes the lamp (Figure 3.11).

Several types of fluorescent lamps are in use, as shown in Figure 3.12. The standard (430 mA), medium bi-pin is the most common. Available in lengths to 48 in., standard bi-pin lamps are low in wattage, a 4-ft lamp burning 40 W. "White" lamps include cool white, warm white, and daylight.

Recessed double-contact lamps are higher in wattages and come in lengths up to 96 in. Fluorescent lighting using either of the lamp types described above is in common use in commercial and public buildings.

Where higher light intensity is desired, power-groove lamps may be used. Because the lamp is not smooth on the inside, a wavy path of light is produced. An 8-ft power-groove lamp has a light emission equivalent to a straight tube that is 9 ft long. Thus, more light is generated from the same number of lamps, resulting in lower cost per footcandle. Wattages range from 110W for a 48-in. lamp to 215 W for one 96 in. long.

Moduline lamps are U-shaped, 22½ in. long for a 40-W lamp. They will provide more light on the same area than two 20-W lamps placed side by side.

Double-contact lamps, which are more efficient than standard lamps, are also available. Designated high output (HO), very high output (VHO), and super high output (SHO), they generate more heat and light and are more expensive to operate than standard lamps. They are not normally used in plant maintenance situations.

Self-ballasted fluorescent lamps to replace incandescent lamps in ordinary sockets are also available.

Plant growth fluorescent lamps have been developed over the past 30 years.

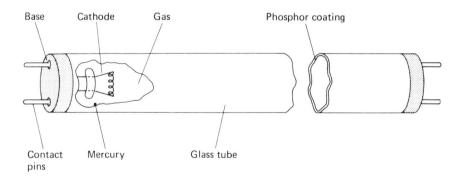

Figure 3.11   Construction of a typical fluorescent lamp.

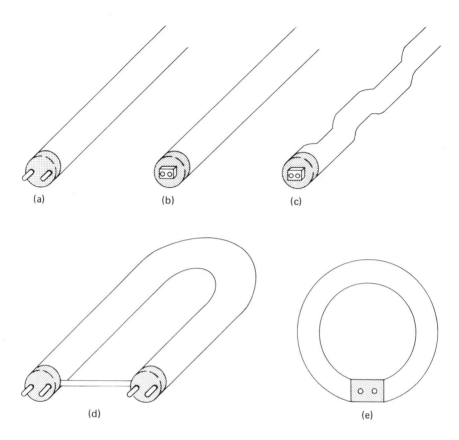

Figure 3.12 Types of fluorescent lamps: (a) standard bi-pin; (b) recessed double contact; (c) power groove; (d) U-shape; (e) circline.

They are purported to provide the proper balance of blue and red light needed by plants. These lamps may be used in the same fixtures as those designed for standard tubes of the same wattage. Some plant growth lamps are reddish blue in color and will distort many colors. They are not satisfactory for general room illumination or reading. Those which exhibit more natural color should no longer be considered plant growth lamps, as yellow-green light has been increased at the expense of red and blue. Plant growth lamps are more expensive than standard lamps of similar wattage and may not be superior to cool white or other "white" lamps.

La Croix et al. (1966) evaluated the effect of three fluorescent lamps with and without incandescent on 11 plant species. For the long term, cool-white fluorescent produced the best results. Gro-Lux lamps were not superior to cool white or warm white, and in many cases were inferior. According to Biran and Kofranek (1976), cool-white fluorescent produced the highest yields with *Tradescantia*. In tests on dry-matter production of nine foliage plants, cool-white lamps had a 9% advantage over the plant growth lamp used. Cathey et al. (1978) tested the effects of seven different fluorescent lamps on 11 home and garden plants. All lamps grew acceptable plants. They suggest that lamp efficiency should be the most important factor in selecting a lamp for plants, with light quality a secondary consideration.

Standard Gro-Lux was the first plant growth lamp and has been available since 1961. The S.E.D. of Gro-Lux closely follows that of chlorophyll synthesis, with peaks in the red and blue (Figure 3.13). Because of its reddish color, Gro-Lux lamps exaggerate the reds and pinks in flowers or foliage.

Gro-Lux Wide Spectrum (WS) lamps have a different S.E.D. from that of Gro-Lux: 8.1% of the light is far-red (735 nm), resulting in a more balanced emission. The presence of far-red light should preclude the need for additional incandescent light when using Gro-Lux WS lamps.

There are other plant growth lamps available. General Electric Plant Lite has an S.E.D. similar to Gro-Lux and is recommended for use in heightening the decorative aspects of plants. Duro-lite produces Vitalite, a full-spectrum lamp purported to match outdoor light in the visible and ultraviolet regions of the spectrum. Verilux Tru-Bloom has a daylight (white) look and provides an even distribution of the visible spectrum for plant growth and general illumination.

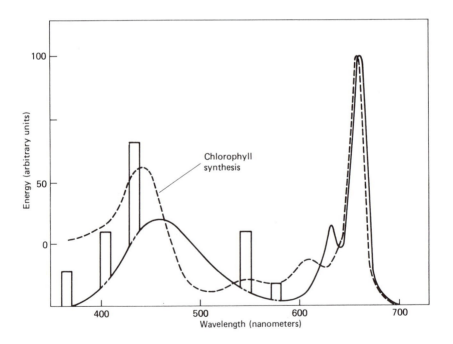

Figure 3.13   Special energy distribution for standard Gro-Lux fluorescent lamps compared to the energy used in chlorophyll synthesis. (Courtesy of GTE Lighting Products.)

Fluorescent lamps require special equipment and complex circuitry. Components must be compatible, with lamps of the same wattage being interchangeable. Commercial or industrial fixtures may be used or, where higher intensities are needed, channels. Ballasts are required. Fluorescent lamps have a high resistance at turn-on, which diminishes rapidly. The ballast limits the current flow; otherwise, the lamp would be destroyed. A good ballast will last 10 to 12 years, with lamp failure, leaking, and smoking an indication of ballast failure and the need for replacement. Ballasts generate heat and must be mounted on metal to reduce the potential of fire.

There are many advantages to using fluorescent lamps for plant maintenance.

1. They are one of the most efficient light generators in common usage, radiating about 22% of the input wattage.

2. They burn cool, with about 36% of the input wattage emitted as infrared heat.

3. Light emission is not changed if dimmers and special ballasts are used.

4. The rated life is 12 to 18 times that of an incandescent lamp.

5. Light distribution is not concentrated at one point and is more even in comparison with incandescent.

6. There is spectral flexibility, depending on the phosphors used.

Disadvantages associated with the use of fluorescent lamps include:

1. Variations in light output are caused by:

   a. *Line voltage*. Lower voltages can cause hard starting and reduce light output. Higher voltages produce higher light output, but light output maintenance is poor and lamp life is shortened.

   b. *Ballast quality*. A poor-quality ballast may not function properly.

   c. *Number and frequency of starts*. The cathodes erode with each start and with frequent starts, thus shortening lamp life.

   d. *Ambient temperature*. Fluorescent lamps operate best at 70 to 90°F, with temperatures above or below this range reducing light output and changing the S.E.D.

   e. *Air movement*. Influences relative humidity and ambient temperature.

   f. *Humidity*. High humidity generally produces hard starting.

   g. *Hours of burning*. Output decreases in time with the greatest change in the first 100 hrs when the lamp is stabilizing.

2. Installation is expensive. Fixtures or components cost more than incandescent.

3. Lamps are low in far-red light.

4. Lamps are not adaptable to situations where focused lighting is needed.

As with all lamps, illumination is reduced with time, and it may be desirable to change the lamps before they burn out. Changing fluorescent lamps after 8,000 hrs of burning is recommended. In many commercial situations, all the lamps are changed periodically, as it is more economical to do so than to replace burned-out lamps one by one. Tubes should be cleaned monthly, if possible.

Long tubes are more efficient than short, as the last 3 in. of the lamp produces very little light Thus, two 48-in. tubes end to end produce less light than one tube 96 in. long. When installing fluorescent illumination, use the longest lamps possible. Also, it is desirable to use two or more lamps side by side rather than single lamps, as the light intensity at any distance from the lamp will be greater. The light from two tubes side by side is more than twice that of a single unit, as shown in Table 3.2. A greater area is also illuminated. Reflectors should be used.

For plant maintenance, a precise color balance is not essential. Cool-white lamps are satisfactory, as is the combination of cool white and warm white. The colored plant growth lamps are not suitable for general illumination, but used in combination with cool white, they will provide good light for both plants and people.

A combination of fluorescent and incandescent lamps is frequently recommended for growth, but is not usually necessary for maintenance. Fixtures are expensive, and the spectral balance is not essential for maintenance. Where this method is used, 10% of the intensity is from incandescent light, with one incandescent watt per two cool-white fluorescent watts the usual recommendation. The addition of incandescent light causes a loss of greenness, making the foliage look slightly yellow.

In retail plant shops and most commercial and public buildings, fluorescent lighting is an efficient method of illumination. Plants will respond favorably to fluorescent light of sufficient intensity and duration.

Mercury Lamps.   Mercury lamps were introduced over 30 years ago. A mercury lamp consists of two glass envelopes, the inner generating the light by passage of an electric arc through mercury vapor, while the outer shields the arc tube from air movement and thermal shock, as well as providing a surface for phosphors and reflectors.

Mercury lamps are available in 40- to 1000-W sizes and may be reflectorized. They require special fixtures and must be ballasted. As with fluorescent lamps, compatible fixtures, ballasts, and lamps must be used.

The energy conversion of mercury lamps is slightly less than that of fluorescent but two to three times greater than incandescent. Visible light is about 13% of the emission and heat, 62%. Lamp life exceeds 24,000 hrs, making these lamps useful in inaccessible fixtures. Lamps are not sensitive to ambient temperature or air movement. As with other lamps, light output declines with lamp age. There is less spectral flexibility with mercury lamps than with fluorescent. The "cold white" of mercury lamps may affect the ambiance of the site; however, color-improved lamps (deluxe white) emit more red light than clear bulbs emit and create a "warmer" atmosphere.

Metal Halide Lamps.   Metal halide lamps are similar to mercury lamps, but have a different S.E.D., owing to the nature of metallic additives to the mercury and a rare gas in the arc tube. The additives are metallic halides (usually iodides) of such metals as thorium, thallium, and sodium, which vaporize, causing the metallic vapors to produce the characteristic spectrum. Different ballasting is also required. Energy conversion is better than mercury or fluorescent, with 20 to 23% visible light and 47% heat. Color uniformity is consistent from lamp to lamp, with a choice of different halides permitting spectral flexibility. Metal halide lamps are available in 175- to 1500-W sizes. Voltage variation will change the S.E.D. Minor changes in color and intensity may also be associated with the burning position of the lamp. Light output declines 30% or more over the life of the lamps, making it necessary to install about 25% more lamps to assure sufficient illumination.

Sodium Lamps. Sodium lamps are a recent introduction. There are two types, high pressure and low pressure. High-pressure sodium (HPS) lamps require a high starting voltage and have an extremely large ballast compared to lamp size. The lamps are extremely efficient, emitting 25 to 27% visible light, most of which is in the 550- to 625-nm (red) range. Available wattages range from 35 to 1000. Spectral flexibility is lacking with the lamps emitting a golden white light. Rated life is 20,000 or more hours.

Low-pressure sodium (LPS) lamps emit 31 to 35% usable light and are the most efficient lamps available. They are available in wattages ranging from 18 to 180 and have a life of 12,000 to 18,000 hrs. The major disadvantage of using LPS is that the light is a monochromatic yellow/orange making the lamps totally unsuited for general lighting situations. LPS may be used in interior plantscapes by lighting when people are not present. Permanent fixtures may be installed, or portable systems positioned over the plants during the period of lighting, and later removed, may be used.

High-intensity discharge (HID) lamps, such as mercury, metal halide, and sodium, have potential in large areas, including lobbies and stores, because they are extremely efficient. Combinations of HID, such as metal halide and HPS, provide good light for general illumination and plants. All require a brief warm-up time to come to full light output. If the power to the lamp is lost or turned off, the arc tube must cool to a given temperature before the arc can be restruck and light produced. The cooling period varies with lamp type, and may be 15 min for metal halide lamps.

Many types of lamps are available for supplemental or total illumination of plants indoors. Table 3.8 compares some of the light sources discussed above. Figure 3.14 shows the S.E.D. for several lamps and that for photosynthesis. If for some reason one wishes to convert from one lamp type to another, the new lamp should emit equal radiant energy. Table 3.9 shows the footcandles of illumination needed from various types of lamps to provide energy equal to cool-white fluorescent. For example, 47 fc from Gro-Lux or 55 fc from summer sun provides radiant energy equal to 100 fc of cool-white fluorescent, and are equally effective for lighting plants.

Incandescent systems are less expensive to install than others, but may be more expensive to maintain, since lamps must be changed more frequently. When one has a choice in selecting an artificial lighting system, the type that will do the job most

TABLE 3.8 Comparison of Light Sources

	Incandescent	Fluorescent	Mercury	Metal Halide	High-pressure Sodium	Low-pressure Sodium
Wattage range	6–1500	4–215	40–1000	175–1500	70–1000	18–180
Lumens per watt	6.5–22.4	24–84	30–63	68–100	80–140	130–180
Lumens per lamp	39–33,620	96–18,000	1200–63,000	11,900–155,000	5800–140,000	2340–32,400
Life (hrs)	750–8000	9000–20,000	16,000–24,000	1500–15,000	20,000–24,000	12,000–18,000
Visible radiation (%)	7–11	22	13	20–23	25–27	31–35
Nonvisible radiation (%)	83	36	62	47–54	47	65–69
Warm-up time (min)	0	0	5–7	2–5	3–4	3–4
Color rendition	Excellent	Good to Excellent	Good	Excellent	Fair	Poor
Concentration	Excellent	Poor	Fair to good	Fair	Fair	Poor
Flexibility	Excellent	Poor	Poor	Poor	Poor	Poor
Lumen output affected by temperature	No	Yes	No	No	No	No
Radio interference	No	Yes	No	No	No	No
Initial cost	Low	Moderate	Moderate	High	High	High
Operating cost	High	Moderate	Moderate	Low	Low	Low

Source: Courtesy of GTE Lighting Products.

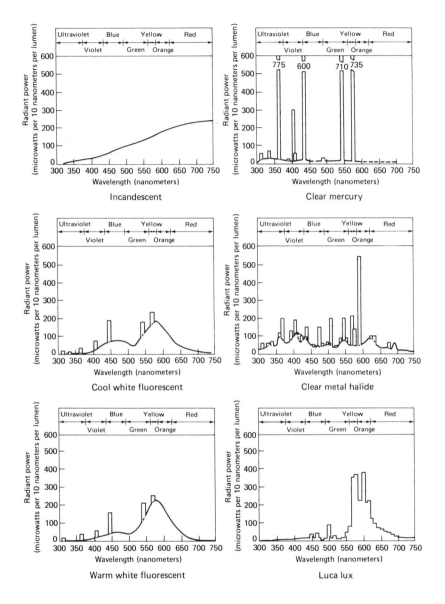

Figure 3.14 (a) Spectral energy distribution for various lamps suitable for interior landscapes; (b) action spectrum for photosynthesis. (Courtesy of General Electric Company.)

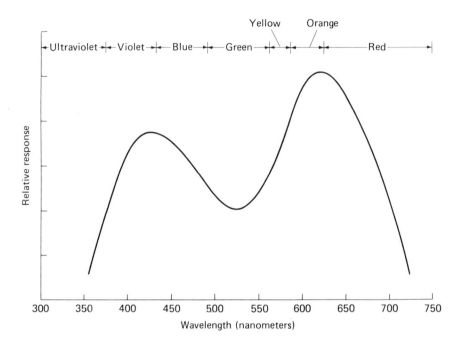

Figure 3.14   (continued)

TABLE 3.9   Footcandles for Equal Radiant Energy (400 to 850 nm) for
Selected Lamps

	Footcandles			
Type of Lamp	100	200	500	1000
Fluorescent				
Cool white	100	200	500	1000
Gro-Lux	47	94	235	470
Gro-Lux WS	68	136	340	680
Agro-lite	74	148	370	740
Vita-Lite	80	160	400	800
Discharge				
Mercury	108	216	540	1080
Metal halide	87	174	435	870
High-pressure sodium	88	176	440	880
Incandescent	35	70	175	350
Sunlight				
Winter	53	106	265	530
Summer	55	110	273	546

Source: H. M. Cathey and L. E. Campbell, *Indoor Gardening,* 1978, p. 20.

effectively and economically should be chosen. Consider lamp efficiency for plant response and human visibility, lamp life, and color rendering. For a few plants, incandescent lamps might be most satisfactory, whereas in larger, low-ceilinged areas, fluorescent lamps are more efficient. In places with high ceilings, HID lighting is best.

Low-Voltage Lighting. Special moods and effects such as highlighting plants or creating shadows and silhouettes can be achieved with low-voltage lighting. In addition, the lighting can provide safety against tripping or slipping, and security and protection against crime.

Sun Lamps. Sun lamps are not intended for plant lighting. They produce high levels of ultraviolet light, more UV at usual distances from plants than plants receive from the sun, and are almost always harmful.

INSTALLATION OF PLANT LIGHTING

Plant lighting should be installed in the area where the plants are located. It is expensive and wasteful of energy to illuminate large areas devoid of plants. In modern offices, task lighting has taken the place of uniform illumination. Adequate light (perhaps 100 fc) is provided over desks, machines, and in other work areas; lower levels, perhaps 35 fc, are used in hallways, lobbies, cafeterias, and so on. In many instances, the plants are to be located in low-light areas. If this is the case, supplemental light is needed. The same is true at home or in the plant shop, where plant lighting must be of sufficient intensity.

How much supplemental light must be added? For best results, a lighting engineer should be consulted. However, knowing the existing light level and the fixture height above the plants, one may calculate the candlepower of the lamp or lamp/fixture combination needed to increase the light intensity to the desired level by using the formula*

$$\frac{\text{candlepower of lamp or}}{\text{lamp/fixture combination}} = \text{footcandles desired} - \text{room footcandles}$$
$$\times \text{ mounting height over the top of the plant, feet}^2$$

For example, what candlepower lamp is needed to provide a minimum of 100 fc if the lamp is 4 ft above the top of the plant in an area with 30 fc of existing light? The formula shows:

$$\text{candlepower} = (100 - 30) \times (4)^2$$
$$= 70 \times 16$$
$$= 1120$$

Four 40-W cool-white fluorescent lamps will satisfy the required lighting needs, assuming some reflectance from the fixture and room surfaces, and minimum

*Adapted from E. D. Bickford, "Choose the Best Light Level," *Florists' Review* 161(4161): 29ff., 1977.

TABLE 3.10   Placement of Plants under Various Lamps for Maximum Growth

Type of Lamp	Low-light Plants (3 W/m² Irradiance)		Medium-light Plants (9 W/m² Irradiance)		High-light Plants (24 W/m² Irradiance)[a]	
	Distance: Plant to Lamp	fc	Distance: Plant to Lamp	fc	Distance: Plant to Lamp	fc
Fluorescent						
Two 40-W, 4-ft lamps						
Cool white	2 ft 7 in.	95	11.75 in.	285	N.S.	N.S.
Warm white	2 ft 7 in.	95	11.75 in.	285	N.S.	N.S.
Gro-Lux Plant Light	2 ft 0 in.	45	7.88 in.	130	N.S.	N.S.
Gro-Lux WS	2 ft 3 in.	65	9.88 in.	190	N.S.	N.S.
Agro-lite	2 ft 3 in.	65	9.88 in.	190	N.S.	N.S.
Vita-Lite	2 ft 3 in.	75	11.75 in.	225	N.S.	N.S.
Two 215-W, 8-ft lamps						
Cool white	5 ft 3 in.	95	2 ft 7 in.	285	11.75 in.	760
Warm white	5 ft 3 in.	95	2 ft 7 in.	292	11.75 in.	760
Gro-Lux Plant Light	4 ft 6 in.	43	2 ft 5 in.	130	9.75 in.	348
Gro-Lux WS	5 ft 0 in.	65	2 ft 3 in.	190	11.75 in.	510
Discharge						
Mercury (400 W)	8 ft 6 in.	100	4 ft 11 in.	300	3 ft 0 in.	800
Metal halide (400 W)	11 ft 6 in.	85	6 ft 6 in.	250	4 ft 0 in.	680
High-pressure sodium (400 W)	13 ft 8 in.	85	8 ft 3 in.	250	5 ft 0 in.	680
Low-pressure sodium (180 W)	6 ft 11 in.	130	3 ft 11 in.	390	2 ft 3 in.	1040
Incandescent						
Standard (100 W)	2 ft 5 in.	32	11.75 in.	97	N.S.	N.S.
PAR38FL (150 W)	11 ft 2 in.	30	6 ft 3 in.	90	4 ft 0 in.	244
Incandescent-mercury (160 W)	9 ft 6 in.	46	5 ft 3 in.	140	3 ft 3 in.	367
Daylight		49		148		400

[a]N.S., not sufficient.

Source: Courtesy of H. M. Cathey.

obstruction from lamp shields. Candlepower information is available from manufacturers of lamps and fixtures.

Another approach that may be useful, particularly at home or in the retail shop, is to first determine from a plant care guide (see Chapter 12) the light requirements for particular plants—low, medium, or high. Using the appropriate columns in Table 3.10, select the style of lamp to be used, and read across to determine the distance the plant should be placed away from the lamp. Moving the plant closer to the lamp increases the light intensity, whereas greater distances cause a reduction. Lamps should burn 8 to 12 hrs daily.

Plant care guides frequently give light requirements in footcandles. Footcandle readings may be changed to watts-per-square-meter irradiance by using the conversions shown in Table 3.11. A value between 0.75 and 3 indicates that the lamp is producing low light. Three to 9 W is medium light, and 9 to 24 W or above is high light. Light intensities at the lower end of each range are sufficient to maintain a specific plant; however, greater satisfaction will be achieved with brighter light.

Plant lighting components should have maximum obscurity. Normally, lighting should be placed directly over the plants, but in certain situations, placement behind or alongside the plants may be desirable. Lighting from one side will cause plants to bend toward the light, necessitating periodic rotation of the pots to keep the growth straight (Figure 3.15). Tilting a fixture reduces light intensity. For example, if a fixture is tilted 30°, light intensity is reduced 14%. A 50% reduction occurs at a 60° angle. Uplighting is not recommended, as leaves are adapted for utilization of light from the top. Photosynthesis will be less efficient, and in time the leaves may turn over if the light comes from below. In addition, uplighting may shine in the eyes of people in the vicinity.

TABLE 3.11   Conversion from Footcandles to Watts-Per-Square-Meter Irradiance

Type of Lamp	Multiply Footcandles By:
Fluorescent	
Cool white	0.030
Warm white	0.030
Gro-Lux Plant Light	0.069
Gro-Lux WS, Agro-lite, Vita-Lite	0.044
Discharge	
Mercury	0.030
Metal halide	0.034
High-pressure sodium	0.034
Low-pressure sodium	0.022
Incandescent	
Standard	0.090
Incandescent-mercury	0.070
Daylight	0.055

Source: Courtesy of H. M. Cathey.

Figure 3.15 Light from one side causes plants to bend toward the light.

All lamps should have a reflector to direct the light at the interiorscaped area with a minimum of loss. These may be built into the lamps, a part of the fixture, or in the form of a reflective surface such as reflective white paint, polished aluminum, and mirrors. When using fluorescent lamps, channel fixtures may be used with standard, nonreflectorized tubes. Glare should be eliminated by the use of valences, parawedge louvers, or fiberglass diffusers. The latter will reduce the light intensity significantly and may require use of additional lamps. Adequate wiring is essential.

All lighting systems should be controlled by electric timers (Figure 3.16).

Figure 3.16 Suitable timers for regulating the hours of lighting.

Simple clocks are adequate and can be set to turn the lights on and off at designated times. Astrological clocks are also available and may be used where natural light is a significant portion of the illumination. They adjust the lighting cycle automatically with the seasonal changes in length of day. To retain quality, plants must produce at least as much food each day as they use. If the light intensity is within the desired range, burning the lamps 12 to 14 hrs every day, 7 days a week, should achieve this objective. Since the effect of the illumination is cumulative, lower light intensities may be compensated for with longer hours of illumination. Twenty-four hours of light has been shown to be harmful to some plant species, including *Brassaia actinophylla, Chamaedorea elegans, Dieffenbachia maculata* 'Perfection', *Dracaena marginata,* and *Ficus benjamina.* Decline in plant quality appears as foliar chlorosis, sometimes followed by necrotic spotting. Conversely, high illumination, as from natural light, reduces the hours of supplemental lighting required. In some situations, a conference room for example, lighting may be accomplished when the room is not in use. Higher intensities for shorter periods may be used in such instances. For plant shops, 150 to 200 fc of light for 12 hrs every day, including Saturdays, Sundays, and holidays, should be adequate.

LIGHT DURATION

The response of a plant to the length of the day is called photoperiodism. Natural day length varies with location on the earth and ranges from 0 to 24 hrs. At 40°N latitude (Philadelphia), the shortest day, December 20, has about 9 hrs of light, while June 20, the longest, has approximately 15 hrs. Three general plant responses to day length have been recognized, short-day, long-day, and indeterminate. Short-day plants respond when the day length is a certain critical number of hours or less; long-day plants require light for a length of time in excess of a critical value; and indeterminate-day plants are not affected by day length. Photoperiodism is controlled by a red/far-red reaction and is normally associated with flowering and reproduction. Other plant responses may also be influenced by length of the day, however.

Day Length and Vegetative Response.   Plant grow better when exposed to long days (Figure 3.17, Table 3.12). In a given light intensity, more photosynthesis occurs with extended light periods, so that more food is available for plant growth and development. Once the needs for maintenance have been satisfied, additional supplies can be stored or used for growth. As previously discussed, as long as the light intensity is above the compensation point, low light for long hours can equal bright light for fewer hours in terms of net photosynthesis.

Long days will reduce the abscission (dropping) of leaves, and delay, or prevent, the onset of the rest period.

Production of plantlets (Figure 3.18) along the leaf margins of certain species of *Kalanchoe* is a long-day response, as is the development of runners of *Saxifraga.* Early production of stolons (spiders) in variegated *Chlorophytum* is induced by a day length of less than 12 hrs, with 3 weeks of 8-hr days resulting in spidering in the

Figure 3.17 Effect of day length and light intensity on growth and flowering of African violet. (Courtesy of General Electric Company.)

TABLE 3.12   Growth of Leaves and Flower Stalks and Flower Production (Average Number) in African Violet at 600 fc in Three Day Lengths

Plant Part	Length of Day		
	6 hrs	12 hrs	18 hrs
Leaves	44.6	54.3	55.7
Flower stalks	18.9	22.6	28.3
Flowers	92.0	180.8	239.0

Source: Courtesy of General Electric Company.

shortest period of time (Figure 3.19). In green-leaved spider plants, long days stimulate greater and more rapid stolon formation. In contrast to Kalanchoe and Chlorophytum, production of plantlets on the leaves of Tolmiea, the piggy-back plant, is a nonphotoperiodic response.

Photoperiodism and Flowering. Flowering is the most frequently recognized aspect of day-length response in plants. In the interior plantscape, one is not usually concerned with the induction of flowering. Some examples of plants responding to photoperiodic induction of flowering are Dendranthema grandiflorum (Chrysanthemum morifolium), Euphorbia pulcherrima (poinsettia), Kalanchoe blossfeldiana, and Gardenia jasminoides which are short-day plants, and tuberous begonias and Petunia, which are long-day plants. Impatiens, Fuchsia, and Begonia semperflorens (wax begonia) are nonphotoperiodic for the flowering response.

Photoperiod may be manipulated to cause the desired plant response. It is a low-energy reaction involving red and far-red light and occurring at light intensities of 1 to 10 fc. Phytochrome is the pigment in the plant perceiving either red or far-red light. Extending the day length will cause a long-day plant to respond and

Figure 3.18   Influence of day length on the production of plantlets in Kalanchoe tubiflora: (left) short day; (right) long day.

(a)

(b)

Figure 3.19 *Chlorophytum* grown under (a) 8-hr day length and (b) 14- and 18-hr day lengths. [From P. A. Hammer and G. Holton: "Asexual Reproduction of Spider Plant, *Chlorophytum elatum,* by Day Length Control," *Florists' Review* 157(4057): 35, 76, 1975.]

inhibit a short-day plant. Conversely, shortening the day induces the desired response in short-day plants while inhibiting long-day plants.

The terms "long-day" and "short-day" are really incorrect because the plant responds to the length of uninterrupted darkness. A short-day plant requires a long, uninterrupted dark period, and even brief periods of light will inhibit the flowering response. The technique for manipulating day length is described in detail in green-house management books.

LIGHT QUALITY

Throughout this chapter, reference has been made to the color of the light and its effect on plants. Visible light when passed through a prism is shown to be composed of several colors, ranging from violet to red. To the human eye, the light appears white. For plants, white light is best. Although most photochemical responses in plants occur in the red/far-red and blue portions of the spectrum, green and yellow light are also physiologically active. A considerable portion of yellow-green light is reflected from leaves, hence their green color. To induce a response, light must be absorbed by one of many plant pigments, such as chlorophyll and phytochrome. A regular sequence of events occurs following absorption, which results in the plant's normal growth pattern.

Responses to Red and Far-Red Light.   Photosynthesis requires red light, with chlorophyll, the green pigment, being the light absorber. When grown in the dark, plants appear white or yellowish and are devoid of chlorophyll, as red light is need-ed for chlorophyll synthesis. Phytochrome is the pigment associated with certain red/far-red light responses, including photoperiodism and other photomorphogenic responses. Seed germination, for example, may depend upon the presence or absence of red or far-red light. Onset of the rest period in many plants is a response to changes in the day length, with short days inducing dormancy. Phytochrome "tells" the plant the relative length of darkness. Red light also promotes stem elon-gation, and when grown solely in red light, most plants will be tall and spindly with wide internodes (spaces between the leaves). Production of anthocyanin, the red pigment found in the vacuoles of certain cells, such as those in the leaves of *Coleus*, *Zebrina*, croton, and red-leaved *Begonia* is enhanced by bright, red light.

Responses to Blue Light.   Blue light, together with red, is essential for maximum photosynthesis and chlorophyll synthesis. Phototropism, the bending of stems toward and roots away from light, is a blue-light response. Generally speaking, plants grown in blue light will be compact, with dark green leaves and thick stems.

SUMMARY

Light is a major determinant of plant growth and development. Among the roles of light are photosynthesis, chlorophyll synthesis, control of stomatal action, photomor-phogenesis, anthocyanin synthesis, and phototropism. In addition, light influences temperature, translocation, mineral absorption, leaf abscission, and transpiration.

Plants respond to three aspects of the light: intensity, duration, and color.

Light intensity is the limiting factor in growing plants indoors. It may be uni-form or variable, depending upon the source of the light and any obstructions or reflectors that may be present. Intensity may be readily measured with either a direct-reading or a photographic light meter. There is a minimum light intensity for each plant and one may select plants adapted to the given light intensity or modify the light intensity to suit the plants.

Light for the interior landscape may come entirely from the sun, from artificial sources (lamps), or be a combination of the two. To save energy, as much natural light as possible should be used. Lamps may serve both general and plant lighting and need not duplicate either the intensity or the color of sunlight. The light must suit both plants and people. "White" light is usually best, providing a good ratio of blue and red/far-red light for plants, while not distorting complexions, plant color, and appearance of furnishings. For plant maintenance, tungsten-filament incandescent and gas discharge lamps such as fluorescent, mercury, metal halide, and sodium are satisfactory.

Incandescent lamps are high in red light. They are easily and inexpensively installed; however, they are not energy-efficient and produce large amounts of heat. These lamps are short-lived.

Fluorescent lamps are high in blue but low in red light. They are one of the most efficient light generators in common usage. They burn cool, and have a rated life 12 to 18 times longer than incandescent, but require special circuitry and are more expensive to install. Plant growth lamps are available, but they may not be superior to standard cool-white or other white lamps.

High-pressure sodium lamps emit 25 to 27% visible light, mostly in the red range of the spectrum. Rated life exceeds 20,000 hrs. Low-pressure sodium lamps are the most efficient available.

Plant lighting should be installed in the area over the plants, and their components should be obscure. Reflectors and timers should be provided. Light must be of sufficient intensity and burn for 12 to 14 hrs daily every day of the week.

Photoperiodism is the response of the plant to length of day. Plants may be designated short-day, long-day, or indeterminate-day. More photosynthesis occurs in long days. Leaf abscission is inhibited. Plantlet production in *Kalanchoe* is a long-day response, with "spidering" on variegated *Chlorophytum* occurring more rapidly in short days. Flowering is the most frequently recognized aspect of photoperiodism.

The color of the light influences plant responses. White light is best. Red and far-red light are important in photosynthesis, chlorophyll synthesis, photoperiodism, onset of dormancy, anthocyanin formation, and elongation of stems. Blue light is used in photosynthesis and chlorophyll synthesis, induces phototropic responses, and produces compact plants.

REFERENCES

Agro-Lite Fluorescent Lamps—Test Results, Westinghouse Lamps, Bloomfield, N.J.

BAUMGARDT, J. P.: "Light for Indoor Plants," *Grounds Maintenance* 12(5): 40–42, 1977.

BICKFORD, E. D.: "Choose the Best Light Level," *Florists' Review* 161(4161): 29ff., 1977.

BICKFORD, E. D., and S. DUNN: *Lighting for Plant Growth*, The Kent State University Press, Kent, Ohio, 1972.

BIRAN, I., and A. M. KOFRANEK: "Evaluation of Fluorescent Lamps as an Energy Source for Plant Growth," *Journal of the American Society for Horticultural Science* 101(6): 625–628, 1976.

BOHNING, R. H., and C. A. BURNSIDE: "The Effect of Light Intensity on Rate of Apparent Photosynthesis in Leaves of Sun and Shade Plants," *American Journal of Botany* 43(8): 557–561, 1956.

BRASWELL, J. H., T. M. BLESSINGTON, and J. A. PRICE: "Influence of Production and Postharvest Light Levels on the Interior Performance of Two Species of Scheffleras," *HortScience* 17(1): 48–50, 1982.

CAMPBELL, L. E., R. W. THIMIJAN, and H. M. CATHEY: "Spectral Radiant Power of Lamps Used in Horticulture," *Transactions of the ASAE* 18(5): 952–956, 1975.

CATHEY, H. M., and L. E. CAMPBELL: *Indoor Gardening*, House and Garden Bulletin 220, U.S. Dept. of Agriculture, Agricultural Research Service, Washington, D. C., 1978.

____: "Light and Lighting Systems for Horticultural Plants," in *Horticultural Reviews*, Vol. 2, J. Janick, ed., AVI Publishing Company, Westport, Conn., 1980, pp. 491–537.

____: "Lighting Viewpoints," *Interior Landscape Industry* 1(11): 22–27, 1984.

____: "Light Sources for Interior Plants," *Interior Landscape Industry* 1(4): 34–39, 1984.

____: "Relative Efficiency of High- and Low-Pressure Sodium and Incandescent Filament Lamps Used to Supplement Natural Winter Light in Greenhouses," *Journal of the American Society for Horticultural Science* 104(6): 812–825, 1979.

____: "Zero-Based Budgeting for Lighting Plants," *American Horticulturist* 57(6): 24–25, 29, 1978.

____, and R. W. THIMIJAN: "Comparative Development of 11 Plants Grown under Various Fluorescent Lamps and Different Durations of Irradiation with and without Additional Incandescent Lighting," *Journal of the American Society for Horticultural Science* 103(6): 781–791, 1978.

COLLINS, P. C., and T. M. BLESSINGTON: "Postharvest Effects of Various Light Sources and Duration on Keeping Quality of *Ficus benjamina* L.," *HortScience* 17(6): 908–909, 1982.

CONOVER, C.A.: "Effects of Acclimatization," *Proceedings of Environmental Conditioning Symposium*, pp. 1–6, Horticultural Research Institute, Inc., Washington, D.C., 1977.

____, R. T. POOLE, and T. A. NELL: "Influence of Intensity and Duration of Cool White Fluorescent Lighting and Fertilizer on Growth and Quality of Foliage Plants," *Journal of the American Society for Horticultural Science* 107(5): 817–822, 1982.

COPLEY, K.: "Light," *Grounds Maintenance* 12(10): 15–21, 1977.

DOWNS, R. J.: "Lighting for Plants and People," Lecture, 1978 National Tropical Foliage Short Course, Orlando, Fla.

DUNN, S.: "Lighting for Plant Growth or Maintenance," *Florists' Review* 156(4054): 41ff., 1975.

FAILS, B. S., A. J. LEWIS, and J. A. BARDEN: "Light Acclimatization Potential of *Ficus benjamina*," *Journal of the American Society for Horticultural Science* 107(5): 762–766, 1982.

FALK, N. K.: "Light Up," *Greenhouse Manager* 4(6): 89–90ff., 1985.

FONTENO, W. C., and E. L. McWILLIAMS: "Light Compensation Points and Acclimatization of Four Tropical Foliage Plants," *Journal of the American Society for Horticultural Science* 103(1): 52–56, 1978.

GAINS, R. L.: *Interior Plantscaping*, Architectural Record Books, New York, 1977.

____: "Trends in Interiors—An Architect's View," *Proceedings of the 1977 National Tropical Foliage Short Course*, pp. 57–62.

Getting the Most From Your Lighting Dollar, The National Lighting Bureau, Suite 300, 2101 L Street N.W., Washington, D.C. 20037.

GRAVES, W. R., and R. J. GLADON: "Ficus and Leaf Drop—An Update," *Interior Landscape Industry* 2(3): 61–64, 1985.

Growing Plants under Fluorescent Light, Union Electric Company, St. Louis, Mo.

Guide to Indoor Garden Lighting with Sylvania Gro-Lux Lamps, GTE Sylvania, Danvers, Mass.

HAMMER, P. A.: "Stolon Formation in *Chlorophytum*," *HortScience* 11(6): 570–572, 1976.

____, and G. HOLTON: "Asexual Reproduction of Spider Plant, *Chlorophtum elatum*, by Day Length Control," *Florists' Review* 157(4057): 35ff., 1975.

HARTLEY, D. E.: "Light Requirements for Foliage Plants," *Foliage Digest* 4(11): 12–13, 16, 1981.

HEINS, R. D., and H. F. WILKINS: "Influence of Photoperiod and Light Quality on Stolon Formation and Flowering of *Chlorophytum comosum* (Thumb.) Jacques," *Journal of the American Society for Horticultural Science* 103(5): 687–689, 1978.

HENLEY, R. W.: "Selection and Use of Light Meters in Nurseries and Indoors," *Foliage Digest* 4(11): 8–10, 1981.

Indoor Gardening with Artificial Light, Duro-lite Lamps, Inc., Fair Lawn, N.J.

KOFRANEK, A. M.: "The Influence of Fluorescent Light on Foliage Plants," *Proceedings of Environmental Conditioning Symposium*, pp. 6–10, Horticultural Research Institute, Inc., Washington, D.C., 1977.

_____: "The Maintenance of Some Indoor Foliage Plants under Fluorescent Light," *Florists' Review* 150(3895): 19–20, 1972.

KOTRAS, J.: "Interior Landscape Lighting," *Southern Florist and Nurseryman* 94(42): 43–44, 1982.

LA CROIX, L. J., D. T. CANVIN, and J. WALKER: "An Evaluation of Three Fluorescent Lamps as Sources of Light for Plant Growth," *Proceedings of the American Society for Horticultural Science* 89: 714–722, 1966.

"Landscape Design Goes Indoors with Plants," *Grounds Maintenance* 2(4): 29–31, 1967.

"Light Plants Efficiently (and Keep Energy Costs Down)," *Florist* 12(4): 53–56, 1978.

MPELKAS, C. C.: *Light Sources for Horticultural Lighting*, Engineering Bulletin 0-352, Sylvania Lighting Center, Danvers, Mass., 1981.

NELL, T. A., C. R. JOHNSON, and J. A. LAURITIS: "Influence of Water Stress and Light Level on Growth, Photosynthesis and Leaf Ultrastructure in *Ficus benjamina*" (abstract), *HortScience* 16(3): 446, 1981.

NORTON, R. A.: "The Present and Future of Artificial Lighting for Increased Plant Growth," *Florists' Review* 153(3955): 65 ff., 1973.

PASS, R. G., and D. E. HARTLEY: "Net Photosynthesis of Three Foliage Plants under Low Irradiation Levels," *Journal of the American Society for Horticultural Science* 104(6): 745–748, 1979.

PETERSON, N. C., and T. M. BLESSINGTON: "Postharvest Effects of Dark Storage and Light Source on Keeping Quality of *Ficus benjamina* L.," *HortScience* 16(5): 681–682, 1981.

Plant Growth Lighting, TP-127, General Electric, Nela Park, Cleveland, Ohio.

PRITCHARD, M. D. W.: *Lighting*, American Elsevier Publishing Company, Inc., New York, 1969.

A Revolution in Plant Growth Lighting, Verilux, Inc., Greenwich, Conn.

Rose Lighting with Lumalux High Pressure Sodium Lamps, GTE Sylvania Engineering Bulletin 0-351, GTE Sylvania, Danvers, Mass.

SALISBURY, F. B., and C. W. ROSS: *Plant Physiology*, Wadsworth Publishing Company, Inc., Belmont, Calif., 1978.

SANDERSON, K.: "Back to Basics—Controlling the Growth of Plants with Light," *Interscape* 5(32): 29ff., 1982.

SCHAUFLER, E. F., and C. C. FISHER: *Artificial Lighting for Decorative Plants*, Extension Bulletin 1087, New York College of Agriculture and Life Sciences, Cornell University, Ithaca, N.Y., 1972.

SPOMER, L. A.: "History and Operation of Energy-Efficient Fluorescent Lamps," *Florists' Review* 167(4329): 20–21, 74, 1980.

STOLZE, J. A. B., ed.: *Application of Growlight in Greenhouses*, Poot Lichtenergie B. V. Schipluiden, Holland, 1985.

THAMES, G., and M. R. HARRISON: *Foliage Plants for Interiors*, Bulletin 327-A, Cooperative Extension Service, Cook College, Rutgers University, New Brunswick, N.J., 1966.

Tungsten Halogen Lamps, Engineering Bulletin 0-349, The Sylvania Lighting Center, Danvers, Mass.

VLAHOS, J., and J. W. BOODLEY: "Acclimatization of *Brassaia actinophylla* and *Ficus nitida* to Interior Environmental Conditions," *Florists' Review* 154(3989): 18ff., 1974.

WEILER, T. C., G. M. PIERCEALL, and J. A. WATSON: "Interior Plantings: The Interior Environment," *Plants and the Landscape* 4(4): 1–4, 1981.

"Wonderlite," *News and Views* 20(1): 6–7, American Horticultural Society, 1978.

Temperature

Plant growth and maintenance depend heavily upon both light and temperature. In Chapter 3, the effects of light on plants were discussed. Temperature influences every physiological process in plants and must be carefully controlled to maximize the longevity of high-quality plants in the interior plantscape.

EFFECTS OF TEMPERATURE

Photosynthesis. Within physiological limits, an increase in temperature will increase the rate of photosynthesis unless some other factor, particularly light, is limiting. Temperatures above 90 to 95°F (32 to 35°C) may be detrimental to plant cells, causing photosynthesis to decline.

Respiration. Increasing temperature will increase the rate of respiration, the utilization of stored food. Plants in cool rooms use less food than those in warm areas. During the day, or when the lights are on, plants photosynthesize and make food. They also respire, using some of the photosynthate produced. At night, or when the lights are out, no food is made, but respiration continues. Thus, photosynthesis must occur at a rate in excess of respiration so that sufficient food is produced for the plant's daily needs and a slight excess is available to permit a small amount of new growth to occur. If possible, reducing the interior temperature at night, or when the lights are out, by 5 to 10°F will reduce the rate of respiration and conserve stored food.

Transpiration. There is a direct relationship between temperature and rate of transpiration. As air is warmed, it expands and can hold more water vapor. If no additional water is added, the amount of water vapor present per unit volume of air declines, resulting in a decrease in relative humidity. Since transpiration is a diffusion process, the rate in low humidity with high temperatures will be very rapid. Water will be pulled through the plant from the growing medium, causing drying and the need for more frequent watering. If moisture is deficient, wilting and desiccation (drying) of plant tissue, especially the leaves, may cause injury. Stomates will close, reducing the water-vapor loss but also curtailing photosynthesis. As the air is cooled, the relative humidity will increase.

Breaking of Dormancy.   Many temperate-region plants, or plant parts, have a period of dormancy and will not resume active growth until exposed to a prescribed period of cold (usually < 40°F or 4.5°C). Most temperate trees and shrubs are not suitable for interior plantscaping because they go dormant indoors even though the environment is favorable for continued growth activity. In deciduous plants, partial or total leaf abscission accompanies the onset of dormancy. Exposed to a constant warm temperature, dormancy is not broken and growth cannot resume. Dormancy is rarely a consideration with tropical interior plants.

For the same reason, bulbs of tulips, daffodils, lilies, and other plants will not grow indoors unless exposed to a period of cold. When these flowering plant materials are used in the plantscape, they are forced in greenhouses and installed as blooming plants.

Protein Synthesis.   There is a reduction in protein synthesis at low temperatures. Carbohydrate accumulation is favored as well as the formation of anthocyanin (red pigment).

OPTIMAL TEMPERATURE RANGE

There is no one temperature at which all plants grow best, but rather an optimal range of temperatures for each plant species. For most tropical foliage plants, a temperature range of 65 to 75°F (18 to 24°C) is satisfactory. It is best to consult plant care lists to determine the needs of specific plants. Table 4.1 lists the suitable temperature range for some frequently used green plants.

A 5 to 10°F reduction of temperature at night is desirable but not absolutely essential. In homes, thermostats are usually lowered at night. In an effort to conserve energy, heating and cooling systems in commercial buildings may be turned off at night and on weekends and holidays, resulting in cold temperatures in winter and excessive heat during the summer. These rapid, extreme temperature fluctuations will be harmful to plants. Quality will decline more rapidly, necessitating more frequent replacement. Your client must understand this, and the maintenance agreement should be drawn accordingly. Similarly, plants may be injured from temperature extremes which may occur during transit and delivery or in the holding facility. Areas adjacent to entrances and windows are subject to extremes of temperature, and plants in these places may require protection from excess heat or cold.

TABLE 4.1   Suitable Temperature Ranges for Some Frequently Used Green Plants

Cool temperature: 55–65°F (13–18°C) day, to 45°F (7°C) night

Araucaria	*Hedera*	*Saxifraga*
Asparagus	*Nephrolepis*	*Yucca*
Cactus	*Podocarpus*	*Zebrina*
Crassula	*Sansevieria*	

Warm temperature: 65–75°F (18–24°C) day, 60–65°F (15.5–18°C) night

Aglaonema	Croton	*Philodendron*
Ananas (pineapple)	*Dieffenbachia*	*Pilea*
Aphelandra	*Dracaena*	*Pittosporum*
Aralia	*Epipremnum*	*Plectranthus*
Araucaria	*Fatsia*	*Podocarpus*
Ardisia	*Ficus*	*Sansevieria*
Asparagus	*Fittonia*	*Saxifraga*
Beaucarnea	*Gardenia*	*Senecio*
Begonia	*Gynura*	*Spathiphyllum*
Brassaia	*Hedera*	*Syngonium*
Chlorophytum	*Hoya*	*Tolmiea*
Cissus	*Hypoestes*	*Tradescantia*
Coffea	*Maranta*	*Yucca*
Cordyline	Palms	*Zebrina*
Crassula	*Peperomia*	

Plant species and cultivars vary in their susceptibility to injury due to either low or high temperatures, and will survive short periods when the temperatures are either higher or lower than those considered optimal. However, depending on the speed of the change and its magnitude, some injury may occur. Slow changes tend to be less harmful than rapid changes of the same amount. Symptoms appear in a matter of hours in susceptible plants but may not be apparent for several days, or even weeks, in more tolerant species.

The severity of chilling injury increases with the length of time leaves are exposed to the cold. It can occur on plants exposed to cold temperatures during shipping, especially to northern markets, and on plants placed in cold building interiors.

Cold Temperatures.   Cold may freeze plants, but many tropical and subtropical plants may be injured by temperatures below 40 to 45°F (4.5 to 7°C) with some extremely sensitive species injured at temperatures below 50 to 55°F (10 to 13°C). Injury associated with low, but nonfreezing temperature is called chilling injury.

Symptoms of cold injury include defoliation; water-soaked and/or necrotic spots or margins on leaves; discoloration, bending, and curling of leaves; leaf wilt; infloresence collapse; poor growth; and death of the plant (Figure 4.1). Leaf symptoms usually appear first on older leaves. Although several factors are undoubtedly

Figure 4.1   Cold temperature causes water-soaked and necrotic spots on leaves of *Philodendron scandens* 'oxycardium'.

involved, chilling injury apparently results from chemical changes in cell membranes. A gradual reduction in temperature prevents rapid chemical change; the plant becomes acclimatized and injury is reduced.

Poole and Conover (1983) studied the severity of chilling injury on 19 foliage plants at 35°F (2°C) for 24 and 48 hrs when grown under different cultural regimes. Table 4.2 shows the results. Twelve of the 19 species tested were injured when exposed to 35°F (2°C) for 24 hrs. *Dracaena surculosa* and *Ficus benjamina* were added to the list after a 48-hr exposure. For 10 of the 14 species injured after 24 hrs, the severity of the injury increased following two days of cold. Increasing watering frequency during production increased chilling injury on eight species. Higher fertilizer levels resulted in more injury to six species (Table 4.2).

Aglaonema 'Silver Queen' is especially sensitive to low, nonfreezing temperatures, with chilling injury occurring at temperatures below 55°F (13°C) (Figure 4.2). Other cold-sensitive *Aglaonema* are *A. nitidum* 'Curtissi,' *A.* x 'Fransher' and *A. commutatum* 'Treubii.' 'Marble Queen' *Epipremnum* also exhibits chilling injury. The whiter the variegation, the more severely it reacts to cold temperature. Other cold-sensitive plants include *Coleus*, *Episcia*, *Exacum*, *Hemagraphis*, *Iresine*, palms, *Pilea involucrata*, *Pilea* 'Moon Valley,' *Spathiphyllum*, and *Syngonium*.

McWilliams and Smith (1978) studied chilling injury in three species using conditions that simulated shipping. Each plant exhibited unique and characteristic symptoms (Figure 4.3). *Epipremnum aureum* was killed after exposure to temperatures of 4.5°C for 4 days. After 2 days of chilling, leaves were black to gray, with water-soaked appearance. Necrosis developed within a week.

After 1 week, plants of *Maranta leuconeura erythroneura* showed cold-injury symptoms—wilting and necrotic lesions on the leaves. Plants exposed to 4.5°C for 6 to 8 days died after 8 weeks in a favorable environment. Within 1 day in a greenhouse, leaves of *Aphelandra squarrosa* 'Louisae' exposed to 4 or more days' chilling became flaccid. Leaf tips and interveinal tissue became necrotic after 2 days' exposure. Abscission occurred following 6 to 8 days' chilling. No plants died. *Pilea cadierei* was not injured by the treatments.

TABLE 4.2 Severity of Chilling Injury on Selected Foliage Plants at 35°F for Different Durations When Grown on Different Cultural Regimes[a]

Botanical Name	35°F Temperature		Watering Frequency		Fertilizer Rate	
	24 hr	48 hr	1/Week	2/Week	7 lb/yd³	7 lb/yd³ + 2 lb/100 gal
Aeschynanthus pulcher	+	+	+	+	+	+
Ardisia crenata	–	–	–	–	–	–
Begonia x rex-cultorum	++	++	++	++	+	+++
Codiaeum variegatum 'Gold Dust'	–	–	–	–	–	–
Dieffenbachia amoena 'Hicolor Cream'	++	++	++	+++	++	++
Dracaena deremensis 'Compacta'	+	+++	+	++	+	+++
D. marginata 'Colorama'	+	++	++	++	–	++
D. surculosa	–	+	–	+	–	+
Ficus benjamina	–	+	–	+	+	+
Fittonia verschaffeltii argyroneura	+	+++	+	+++	++	+++
Hedera helix	–	–	–	–	–	–
Hoya carnosa	+	+	+	+	+	+
Peperomia obtusifolia	+	++	+	++	+	+
P. obtusifolia 'Marble'	+	++	+	+	+	+
P. obtusifolia 'Variegata'	+	++	++	++	++	++
Philodendron scandens 'oxycardium'	–	–	–	–	–	–
Plectranthus australis	+	++	+	++	+	+
Sansevieria trifasciata 'Laurentii'	+	+	+	+	+	+
Senecio rowleyanus	–	–	–	–	–	–

[a]Chilling damage ratings: – no damage (no effect due to treatment); + slight damage (damage restricted to less than 10% of leaves); ++ moderate damage (damage restricted to less than 50% of leaves); +++ severe damage (damage on 50% or more of leaves).

Source: R. T. Poole and C. A. Conover, Interscape 5(41): 12–13, 1983.

Figure 4.2 Chilling injury on *Aglaonema commutatum* 'Silver Queen': (left) normal plant; (right) after exposure to 45°F, leaves are pale and variegation is reduced.

In other work, Smith and McWilliams (1978) showed that *Epipremnum* and *Maranta* preconditioned by exposure to 16°C for 9, 18, or 27 days resisted chilling exposures (4.5°C) which killed unhardened plants.

Selection of cold-tolerant plants for areas of low temperature may solve an interior landscape problem. Table 4.3 lists some suitable plants.

High Temperatures.   Heat may be harmful to plants by causing excessive transpiration, which results in wilting and desiccation of tissues. Respiration will increase, causing a depletion of stored food. Growth is stimulated which, in the low light of interiors, will be spindly. Death of the entire plant may result from too-high temperatures. With rapid warming, coagulation of proteins occurs, thereby disrupting protoplasmic structure. When the warming is more gradual, proteins are broken down, releasing ammonia, which is toxic. Buck and Blessington (1982) have shown that severe foliage injury occurs to *Ficus benjamina* and *F. lyrata* when exposed to 37°C (97°F) for 6 or 9 days under conditions simulating shipping. Some plants tolerant of high temperatures for brief periods are listed in Table 4.4.

Measuring Temperature

A thermometer placed near the plants is the best way of assessing temperature (Figure 4.4). Considerable variation will exist in the temperature throughout a room, and remote thermometers, such as room thermostats, will not give accurate readings

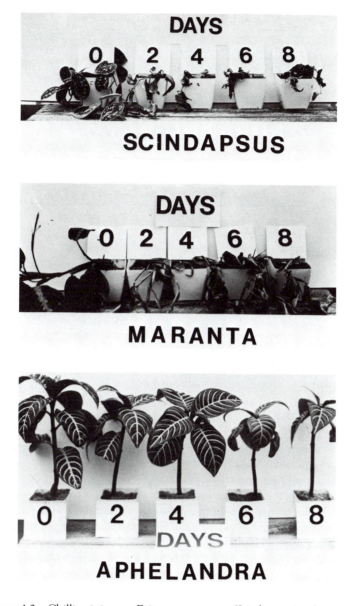

Figure 4.3 Chilling injury on *Epipremnum aureum (Scindapsus pictus)*, *Maranta leuconeura*, and *Aphelandra squarrosa* 2 weeks after exposure to 0, 2, 4, 6, and 8 days at 4.5°C in the dark. [From E. L. McWilliams and C. W. Smith, "Chilling Injury in *Scindapsus pictus, Aphelandra squarrosa*, and *Maranta leuconeura*," *HortScience* 13(2): 179–180, 1978.]

TABLE 4.3   Some Cold-Tolerant Plants

Araucaria	Fatsia
Ardisia	Hedera
Aspidistra	Hoya
Beaucarnea	Pilea cadierei
Brassaia	Pittosporum
Chlorophytum	Podocarpus
Crassula	Saxifraga
Croton 'Gold Dust'	Yucca elephantipes

TABLE 4.4   Some Heat-Tolerant Plants

Aglaonema	Cordyline
Ananas	Dizygotheca
Aphelandra	Dracaena
Aspidistra	Hedera
Beaucarnea	Hoya
Brassaia	Palms
Cactus	Peperomia
Chlorophytum	Succulents

for the specific places in which the plants are located. Periodic readings, or a thermograph, will enable one to determine any fluctuation that may occur.

Some Precautions

When plants are placed near a window, sunlight through the glass quickly warms leaf surfaces and may scorch them. To eliminate the possibility of injury, the plant may be moved or some shading provided.

Cold outside temperatures make the inside air next to the glass cold, particularly at night. Freezing or other plant injury may occur due to the sudden, extreme temperature fluctuation. Avoid, or reduce, the threat of injury by using storm windows, double glazing (Thermopane), caulking, and weatherstripping. Close drapes or pull shades between the plants and the glass at night. Sheets of newspaper taped inside the glass provide insulation. Covering plants at night will protect them, as will moving them to a warmer location away from the glass. Plants should not be placed in front of heating or air-conditioning ducts, or on radiators. Placing plants on a board above a radiator will insulate them from the heat. Not all of these measures are practical in commercial plantings, where consideration of possible temperature extremes should be given initially when locating the plants.

Interior landscapers should be certain that the heating, air-conditioning, and ventilating systems are operating and balanced before any plant material is placed into the area. Not to do so could be disastrous and extremely costly.

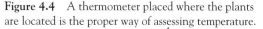

Figure 4.4 A thermometer placed where the plants are located is the proper way of assessing temperature.

Plants should be acclimatized to changes in temperature. At home, if plants are put outdoors during the summer, it should be after the nights are above 60°F (15°C). They should be returned inside before the night temperature drops below 60°F again; otherwise, they will adapt to cooler temperatures and must be acclimatized before placement in the normal room environment. Plants from local greenhouses and southern and western production areas should be acclimatized before using them indoors. Chapter 11 describes the procedures to be used.

SUMMARY

Temperature affects every physiological process in plants. An optimal temperature range exists for each species, with 65 to 75°F (18 to 24°C) suitable for most indoor plants. A daily fluctuation is not necessary. Sudden, excessive fluctuations may be harmful, and plants should be protected from them. Acclimatization to any temperature change is desirable. Thermometers placed near plants accurately assess temperature.

REFERENCES

BUCK, T. L., and T. M. BLESSINGTON: "Postharvest Effects of Temperatures during Simulated Transit on Quality Factors of Two *Ficus* Species," *HortScience* 17(5): 817–819, 1982.

FISHER, C. C., and R. T. FOX: *The Selection, Care, and Use of Plants in the Home*, Information Bulletin 117, New York State College of Agriculture and Life Sciences, Cornell University, Ithaca, N.Y., 1977.

FOOSHEE, W. C. and D. B. MCCONNELL: "Response of *Aglaonema* 'Silver Queen' to Nighttime Chilling Temperatures," *HortScience* 22(2): 254–255, 1987.

HUMMEL, R. L. and R. J. HENNY: "Variation in Sensitivity to Chilling Injury Within the Genus *Aglaonema*," *HortScience* 21(2): 291–293, 1986.

MARLATT, R. B.: "Chilling Injury in *Sansevieria*," *HortScience* 9(6): 539–540, 1974.

_____, and D. B. MCCONNELL: "Chilling Injury of Field-Grown *Sansevieria* in South Florida," *Foliage Digest* 2(10): 7–8, 1979.

MAROUSKY, F.: "Chilling Injury in *Dracaena sanderana* and *Spathiphyllum* 'Clevelandii'," *HortScience* 15(2): 197–198, 1980.

_____: "The Shipping Environment," Lecture, 1978 National Tropical Foliage Short Course, Orlando, Fla.

MCCONNELL, D. B., D. L. INGRAM, C. GROGA-BADA, and T. J. SHEEHAN: "Chilling Injury of Silvernerve Plant," *HortScience* 17(5): 819–820, 1982.

MCWILLIAMS, E. L., and C. W. SMITH: "Chilling Injury in *Scindapsus pictus*, *Aphelandra squarrosa*, and *Maranta leuconeura*," *HortScience* 13(2): 179–180, 1978.

MEYER, B. S., D. B. ANDERSON, R. H. BOHNING, and D. G. FRATIANNE: *Introduction to Plant Physiology*, D. Van Nostrand Company, New York, 1973.

NOGGLE, G. R., and G. J. FRITZ: *Introductory Plant Physiology*, Prentice-Hall, Inc., Englewood Cliffs, N.J., 1976.

POOLE, R. T., and C. A. CONOVER: *Factors Influencing Damage of Foliage Plants*, ARC-A Research Report RH-83-4, University of Florida, IFAS, Agricultural Research Center-Apopka, 1983.

Professional Guide to Green Plants, Florists' Transworld Delivery Association, Southfield, Mich., 1976.

SALISBURY, F. B., and C. W. ROSS: *Plant Physiology*, Wadsworth Publishing Company, Inc., Belmont, Calif., 1978.

SMITH, C. W., and E. L. MCWILLIAMS: "Chilling, Hardening, and Electrolyte Loss in *Scindapsus pictus* and *Maranta leuconeura*" (abstract), *HortScience* 13(3): 344, 1978.

The Atmosphere

The atmosphere consists of the gaseous envelope surrounding the plant and the masses of gas in the growing medium and within plant tissue. The air inside an average building contains about 79% nitrogen (N), 21% oxygen (O_2), 0.03% carbon dioxide (CO_2), other gases, and water vapor, dust and dirt, microorganisms, and pollen. In contrast, the growing medium usually has higher levels of CO_2 and lower levels of O_2 than the ambient air because of the slow rate of gas exchange between the medium and the atmosphere.

OXYGEN AND CARBON DIOXIDE

All living plant cells require O_2 for respiration, the process that releases energy for all plant processes. Respiration occurs 24 hrs a day in all living cells of plants and animals and produces CO_2 in addition to energy. In the presence of light, green plant cells use CO_2 in the process of photosynthesis, to produce glucose, the basic food on which all life depends, plus oxygen.

The O_2/CO_2 content of the air in the average building may deviate slightly from that of the external atmosphere but is adequate to satisfy the physiological needs of the plant.

Although plant cells consume oxygen for respiration 24 hrs a day, there is no danger that they will deplete the supply in a room and cause a health hazard to human beings or pets. For example, a person would have to remain 12 hrs in a completely dark, 800-ft^3 room with no air exchange, and three hundred 2-ft schefflera plants, before suffering from lack of oxygen. Even in a closed room, O_2 depletion will not occur, as forced ventilation in modern commercial buildings, and air seepage and ventilation at home, help maintain proper O_2/CO_2 ratios. Also, since the plants will receive light, they will generate O_2.

In the growing medium, an oxygen deficiency or carbon dioxide excess may occur if drainage is poor or overwatering has occurred. If the fluctuation is sufficient, injury or death of the root system may occur.

DRAFTS

Plants should not be subject to drafts of hot or cold air. Extreme temperatures may injure the plants directly, and the air movement may cause excessive transpiration with potential wilting, injury, and even death of plant tissue (Figure 5.1).

RELATIVE HUMIDITY

The relative humidity (RH) of air is the actual humidity of the air compared to potential humidity (vapor pressure) of the air at a given temperature. For a given amount of moisture, the relative humidity will be higher in cold than in warm air. Heating this air will lower the humidity; cooling it will increase the relative humidity again.

Most of the plants used in interior landscapes have been produced in an environment where the relative humidity ranged from 85 to 95%. This is far in excess of the 40% or lower relative humidity of many building interiors. The relative humid-

Figure 5.1 Chinese evergreen injured by excessive transpiration in a hot room with low relative humidity.

ity of most American homes in winter is usually below 15%, a level considered too low for human health and comfort. The Sahara desert has 25% relative humidity and Death Valley, 23%.

Relative humidity is important because it affects transpiration, hence plant–water relationships. Although most tropical foliage plants thrive at humidities greater than 30%, they will survive in the low-moisture environments of building interiors if they are properly acclimatized.

Measuring Humidity

Measuring the humidity in building interiors is readily accomplished with an instrument called an hygrometer (Figure 5.2). A wet/dry-bulb hygrometer consists of two thermometers: the bulb of one is dry, while the other is covered with a piece of wet gauze. As water evaporates from the gauze, the temperature of the wet-bulb thermometer will be depressed compared with the dry bulb. The lower the relative humidity, the greater the difference in the two readings. Comparison of the ambient temperature with the degrees of depression in Table 5.1 will give the relative humidity. For example, if the air temperature (dry bulb) is 70°F and there is a 15°F difference between the dry and the wet bulb, a relative humidity of 36% is found by

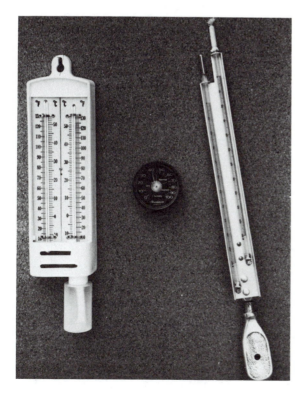

Figure 5.2 Hygrometers: (left) wet/dry-bulb hygrometer; (center) direct reading; (right) psychrometer.

TABLE 5.1 Relative Humidity as Determined by Wet- and Dry-bulb Thermometers

Air Temperature (°F)	Difference between Dry and Wet Thermometers (°F)																																			
	1	2	3	4	5	6	7	8	9	10	11	12	13	14	15	16	17	18	19	20	21	22	23	24	25	26	27	28	29	30	31	32	33	34	35	36
30	89	78	67	56	46	36	26	16	6																											
35	91	81	72	63	54	45	36	27	19	10	2																									
40	92	83	75	68	60	52	45	37	29	22	15	7																								
45	93	86	78	71	64	57	51	44	38	31	25	18	12	6																						
50	93	87	80	74	67	61	55	49	43	38	32	27	21	16	10	5																				
55	94	88	82	76	70	65	59	54	49	43	38	33	28	23	19	14	9	5																		
60	94	89	83	78	73	68	63	58	53	48	43	39	34	30	26	21	17	13	9	5	1															
65	95	90	85	80	75	70	66	61	56	52	48	44	39	35	31	27	24	20	16	12	9	5	2													
70	95	90	86	81	77	72	68	64	59	55	51	48	44	40	36	33	29	25	22	19	15	12	9	6	3											
75	96	91	86	82	78	74	70	66	62	58	54	51	47	44	40	37	34	30	27	24	21	18	15	12	10	7	5	3								
80	96	91	87	83	79	75	72	68	64	61	57	54	50	47	44	41	38	35	32	29	26	23	20	18	15	12	9	7	4	1						
85	96	92	88	84	80	77	73	70	66	63	60	56	53	50	47	44	41	38	36	33	30	28	25	22	20	17	15	13	11	9	6	4	2			
90	96	92	89	85	81	78	74	71	68	65	61	58	55	52	49	47	44	41	39	36	34	31	29	26	24	22	19	17	15	13	11	9	7	5	3	1
95	96	93	89	86	82	79	76	72	69	66	63	60	58	55	52	49	47	44	42	39	37	35	32	30	28	25	23	21	19	17	15	13	11	10	8	6
100	96	93	89	86	83	80	77	73	70	68	65	62	59	56	54	51	49	46	44	41	39	37	35	33	30	28	26	24	22	21	19	16	15	13	12	10
105	97	93	90	87	84	81	78	75	72	69	66	64	61	58	56	53	51	49	46	44	42	40	38	35	33	31	30	28	26	24	22	20	19	17	15	14
110	97	93	90	87	84	81	78	75	73	70	67	65	62	60	57	55	52	50	48	46	44	42	40	38	36	34	32	30	28	26	25	23	21	20	18	17
115	97	94	91	88	85	82	79	76	74	71	69	66	64	61	59	57	54	52	50	48	46	44	42	40	38	36	34	33	31	29	28	26	24	23	21	20
120	97	94	91	88	85	82	80	77	74	72	69	67	65	62	60	58	55	53	51	49	47	45	43	41	40	38	36	34	33	31	29	28	26	25	23	22

Figure 5.3 Electronic hygrometer. (Courtesy of Panametrics.)

locating 70° in the air temperature column and 15° in the "difference" column and following the horizontal and vertical lines, respectively, until they intersect. A sling psychrometer is a modification of a fixed wet/dry-bulb hygrometer. It enables one to swing the two thermometers in the air, thereby producing the temperature differential more rapidly and reducing the time to take the measurement. Direct-reading hygrometers accurate to within ±2% are available as are battery-powered, solid-state electronic hygrometers (Figure 5.3). Hand-held portable units have a liquid-crystal display that reads directly in percent relative humidity. Accuracy is within ±2% over the range 5 to 90% at ambient temperatures between 10 and 50°C. Some units also have temperature measurement capabilities.

Increasing Humidity

Increasing the humidity around indoor plants is desirable, and various methods may be used. Humidifiers installed in central heating and cooling systems humidify entire buildings. This is the only practical way in commercial installations, but if it is not available one must work with existing conditions. In rooms with plants, small humidifying units are suitable. Mist blowers that use centrifugal force to atomize water should be used, not vaporizers, which may become too hot. Humidistats should be used to activate the equipment and maintain the desired moisture level.

Placing plants in areas of high humidity will take advantage of moisture already present. At home, locations above the kitchen sink or in the bathroom may

Figure 5.4 Fountains and pools add to the beauty of the interior plantscape and increase relative humidity.

be ideal for plants. Plants add mineral-free, microbial-free moisture to the indoor environment to raise relative humidity. As water is transpired from plant leaves and evaporates from the surface of the medium, relative humidity is increased. The rate and amount of water added to indoor environments depends upon the plant species, size, moisture content of the growing medium, temperature, and relative humidity inside. The more plants, the greater the change. Setting the plants above a tray filled with gravel, sand, or peat that is kept moist will provide a more humid atmosphere. Do not set the plants in the water, as overwatering may result.

Water accents in the interior garden, such as fountains and pools, are an attractive addition to an extensive planting and reduce the dryness of the atmosphere (Figure 5.4). Closed structures, such as terrariums, greenhouses, atriums, tents, and plastic bags, will also create a humid atmosphere inside. Care must be exercised, however, to ensure that closed areas do not become excessively hot. The recommended practice of misting the plants several times daily is at best a futile attempt to increase the humidity. The brief increase in relative humidity does not sufficiently benefit the plant to justify the practice. Further, it may spread disease organisms.

AIR POLLUTANTS

Sick Building Syndrome

In recent years, a new building term, "sick building," has come into use. A building is said to be "sick" when its occupants complain of health problems that can be related to working or being in a building. These problems are referred to as "sick building syndrome" (SBS), and are generally related to air pollution problems. In some buildings, air may be 100 times more polluted than that outdoors.

Contributing to SBS are gases emitted from furniture, carpets, drapes, copying machines, insulation, paint, and construction materials. Tobacco smoke adds respirable particulates. Biological contaminants such as bacteria, molds and their spores, pollen, and viruses affect air quality. Humans, through normal biological processes, emit hundreds of organic substances. Outside sources such as vehicle exhaust also contribute to indoor air pollution. Reduced ventilation intensifies the problem. Contaminants usually act in combination, and often supplement other occupant complaints such as inadequate temperature, humidity, or lighting.

According to government estimates, there are at least 70 million people who work in nonindustrial buildings—including 27 million office workers—who risk exposure to SBS.

Symptoms of SBS include headache; eye, nose, and throat irritation; respiratory irritation; dry or itchy skin; dizziness and nausea; difficulty in concentrating; fatigue; and sensitivity to odors. Relief occurs upon leaving the building.

A natural solution to SBS problems is plants. NASA initiated research into the potential use of plants as reducers of indoor air pollution on Earth, and in future space habitats. In a 1989 NASA report, researchers B. C. Wolverton, Anne Johnson, and Keith Bounds reported the results of this joint effort of NASA and the Associated Landscape Contractors of America.

In the study, leaves, roots, soil, and associated microorganisms of plants were evaluated. Chemical contamination tests were conducted in Plexiglas chambers on a variety of commonly used indoor plants. Initially, three gases were studied, benzene, formaldehyde, and trichloroethylene. Subsequent studies included ammonia. All are common in the indoor atmosphere and have been shown to produce health problems in people. Results showed that house plants, especially those requiring low light, removed nearly 87% of the pollutants from the chambers within 24 hours. Soil microorganisms have also been shown to degrade toxic organic chemicals with exudates from plant roots significantly influencing the microorganisms that can grow in the root area. Interior plants remove organic chemicals from the indoor atmosphere by a combination of leaf and soil absorption and biodegradation by plant enzymes and/or root-zone microorganisms. It has been suggested that one healthy plant per 100 square feet of floor space is adequate for the average home. Pots should be at least six inches in diameter.

In a subsequent study, Bill and John Wolverton determined which foliage plants were effective in removing formaldehyde, xylene, and ammonia from sealed indoor environments. Tables 5.2 through 5.6 list the ability of indoor plants to remove formaldehyde, benzene, trichloroethylene, xylene, and ammonia, respectively, from the air in sealed chambers.

Wolverton's research and other studies by respected scientists are largely accepted by the scientific community. Plants do clean the air. Since the studies were conducted in sealed chambers, questions about the effectiveness of plants in cleaning the air in an office environment have been raised.

TABLE 5.2   Removal of Formaldehyde from Sealed Chambers by Plants

Ranking	Scientific Name	Common Name	Removal Rate ug/hr
1	Nephrolepis exaltata 'Bostoniensis'	Boston fern	1863
2	Dendranthema x grandiflorum (Chrysanthemum x morifolium)	Chrysanthemum	1454
3	Phoenix roebelenii	Dwarf date palm	1385
4	Dracaena deremensis 'Janet Craig'	Janet Craig	1361
5	Chamaedorea seifritzii	Bamboo palm	1350
6	Nephrolepis obliterata	Kimberley queen fern	1328
7	Hedera helix	English ivy	1120
8	Ficus benjamina	Weeping fig	940
9	Spathiphyllum wallisii 'Clevelandii'	Peace lily	939
10	Chrysalidocarpus lutescens	Areca palm	938
11	Dracaena fragrans 'Massangeana'	Corn plant	938

Source: Adapted from Wolverton, B. C. and J. Wolverton, "Interior Plants and Their Role in Indoor Air Quality: An Update," Interiorscape 11(4): 17, 1993.

TABLE 5.3.   Removal of Benzene from Sealed Chambers by Plants

Ranking	Scientific Name	Common Name	Removal Rate ug/hr
1	Gerbera jamesonii	Gerbera daisy	4485
2	Dendranthema x grandiflorum (Chrysanthemum x morifolium)	Chrysanthemum	3205
3	Spathiphyllum 'Mauna Loa'	Peace lily	1725
4	Dracaena deremensis 'Warneckii'	Striped dracaena	1629
5	Chamaedorea seifritzii	Bamboo palm	1420
6	Dracaena marginata	Dragon tree	1264
7	Sansevieria laurentii	Snake plant	1196
8	Dracaena deremensis 'Janet Craig'	Janet Craig	1082
9	Aglaonema commutatum 'Silver Queen'	Chinese evergreen	604
10	Hedera helix	English ivy	579

Source: Adapted from Wolverton, B. C., A. Johnson, and K. Bounds, Interior Landscape Plants for Indoor Air Pollution Abatement, National Aeronautics and Space Administration, John C. Stennis Space Center, Stennis Space Center, MS, 1989, p. 10.

In October 1994, a three-year study was initiated in Toronto to determine exactly how effective plants are at purifying indoor air and under what conditions. A 1,700-sq-ft conference room was transformed into a tropical ecosystem with 7,500 plants plus mollusks, amphibians, fish, and insects. Accommodating up to 75 people, the ecosystem is exposed daily to numerous synthetic and human contaminants. Air quality testing began in early 1995.

TABLE 5.4.   Removal of Trichloroethylene from Sealed Chambers by Plants

Ranking	Scientific Name	Common Name	Removal Rate ug/hr
1	Gerbera jamesonii	Gerbera daisy	1622
2	Dracaena marginata	Dragon tree	1137
3	Spathiphyllum 'Mauna Loa'	Peace lily	1127
4	Dracaena deremensis 'Janet Craig'	Janet Craig	764
5	Chamaedorea seifritzii	Bamboo palm	688
6	Dracaena deremensis 'Warneckii'	Striped dracaena	573
7	Dracaena fragrans 'Massangeana'	Corn plant	421
8	Sansevieria laurentii	Snake plant	405
9	Hedera helix	English ivy	298

Source: Adapted from Wolverton, B. C., A. Johnson, and K. Bounds, *Interior Landscape Plants for Indoor Air Pollution Abatement,* National Aeronautics and Space Administration, John C. Stennis Space Center, Stennis Space Center, MS, 1989, p. 9.

TABLE 5.5.   Removal of Xylene from Sealed Chambers by Plants

Ranking	Scientific Name	Common Name	Removal Rate ug/hr
1	Chrysalidocarpus lutescens	Areca palm	654
2	Phoenix roebelenii	Dwarf date palm	610
3	Dieffenbachia maculata 'Camille'	Dumb cane	341
4	Dracaena marginata	Dragon tree	338
5	Dieffenbachia maculata	Dumb cane	325
6	Homalomena sp.	King of hearts	325
7	Nephrolepis obliterata	Kimberly queen fern	323
8	Dracaena deremensis 'Warneckii'	Striped dracaena	295
9	Anthurium andraeanum	Lady Jane	276
10	Dracaena fragrans 'Massangeana'	Corn plant	274

Source: Adapted from Wolverton, B. C. and J. Wolverton, "Interior Plants and Their Role in Indoor Air Quality: An Update," *Interiorscape* 11(4): 18, 1993.

Other solutions to sick building problems include pollution source removal, modification, or substitution—for example, cleaning or replacing dirty filters in the HVAC system; removing water-stained ceiling tile and carpeting; and storing paints, adhesives, and solvents in a well-ventilated area. Also, pollutant sources should be used when the fewest people are present: painting should be done on weekends or during nonworking hours, for example. Increasing ventilation rates can also be helpful. This will increase building maintenance costs, and, in many cities, may simply exchange indoor pollutants for outdoor pollutants. In fact, high ventila-

TABLE 5.6.   Removal of Ammonia from Sealed Chambers by Plants

Ranking	Scientific Name	Common Name	Removal Rate ug/hr
1	Rhapis excelsa	Lady palm	7,356
2	Homalomena sp.	King of hearts	5,208
3	Liriope spicata	Lily turf	4,308
4	Anthurium andraeanum	Lady Jane	4,119
5	Dendranthema x grandiflorum (Chrysanthemum x morifolium)	Chrysanthemum	3,641
6	Calathea elliptica 'Vittata'	Peacock plant	3,100
7	Tulipa 'Yellow Present'	Tulip 'Yellow Present'	2,815
8	Chamaedorea elegans	Parlor palm	2,453
9	Ficus benjamina	Weeping fig	1,480
10	Spathiphyllum wallisii 'Clevelandii'	Peace lily	1,269

Source: Adapted from Wolverton, B. C. and J. Wolverton, "Interior Plants and Their Role in Indoor Air Quality: An Update," Interiorscape 11(4): 18, 1993.

tion rates may overwhelm the plant's effectiveness in removing indoor polluting chemicals. Finally, the installation of air filtration and purification systems can do much in combatting sick building syndrome.

Studies by Wolverton, reported by Clein (1993), have shown the effects of foliage plants on airborne microbes and humidity. Airborne microbial levels in a sunroom in which 33% of the surface area was covered with 15 different foliage plant species were significantly lower than in a control room without plants. Humidity levels were 21% higher and airborne levels 65% lower in the sunroom. Wolverton suggests that the plants are emitting substances from their leaves that, mixed with water vapor, suppress the growth of airborne microbes. This helps explain how low-light-requiring foliage plants that evolved in humid environments beneath the canopy of tropical rain forests protect themselves from the invasion of molds, mildew, and other microbes that thrive in damp environments.

Other Pollutants

Ethylene may cause injury to established plantings, and in shipping. A product of hydrocarbon (oil, gas, gasoline, coal) combustion, ethylene may be produced by the heating systems of buildings, and is a product in the exhaust systems of our vehicles. Should ethylene seep into the atmosphere of buildings, plant injury may result.

Plant tissue also produces ethylene. In shipping, plants in closed cartons generate ethylene, which may reach levels sufficient to cause injury. Any fungi that may be present will also contribute to ethylene synthesis. Foliage plants should not be shipped in the same compartments with fruits and vegetables. Apples, pears, peaches, avocados, nectarines, apricots, and cantaloupes have been shown to be high producers of ethylene. Honeydew melons and tomatoes are medium producers, and strawberries, cucumbers, beans, squash, cherries, and citrus, low. Should a truck's exhaust system be malfunctioning, it, too, may contribute to increased ethylene during shipping.

A typical symptom of ethylene injury is epinasty, a response in which the leaves bend downward along the stem. In plants, ethylene is a hormone, and other symptoms of ethylene injury include yellowing of leaves and abscission (dropping) of older or all leaves (Figure 5.5), and growth retardation, either separately or in combination. In flowering plants, abnormal flower development and premature flower drop may occur. Five parts per million of ethylene in the air is harmful to *Aphelandra, Brassaia, Coleus, Crassula, Ficus, Fittonia, Peperomia,* and *Philodendron. Dracaena* and *Hoya* are tolerant to 5 ppm.

Research reported by Marousky and Harbaugh (1979) and Marousky (1979) showed that leaf abscission in *Philodendron* and *Fittonia* increased as the concentration of ethylene and exposure time increased. For the same ethylene concentration, injury was less severe at low temperature (16°C), well above the chilling level, than at 23.5 or 27°C. Leaf abscission occurred regardless of whether exposure to ethylene was in light or dark; however, the species differed in their response. *Philodendron* lost more leaves when exposed in the light, while *Fittonia* was equally affected by ethylene exposure in light or dark.

These results have implications for shippers who must pack and ship plants correctly, and retailers, plantscapers, and others receiving plants shipped long distances. Cartons should be opened immediately upon delivery to provide fresh air.

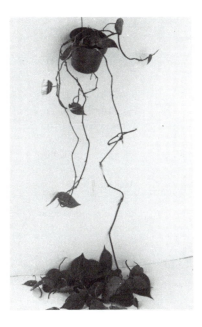

Figure 5.5 Exposure of *Philodendron scandens* 'oxycardium' to ethylene causes leaf abscission.

Remove plant sleeves, especially if they are plastic. Ventilation of rooms will reduce the potential for ethylene toxicity. Adjustment of burners to ensure full combustion and control of exhaust emissions from vehicles will also reduce the ethylene hazard.

Since ethylene exposure can predispose a plant to deterioration, plants may arrive at their destination with no visible symptoms of injury, but develop chlorosis and abscise leaves in a few days.

Atmospheric fluoride, a product of industrial processes, has been shown to adversely affect foliage plants. Table 5.7 lists the susceptibility of various plants that may be used indoors. According to Woltz and Waters (1978), the pattern of fluoride toxicity symptoms appears related to the passive movement of water with the transpirational stream, resulting in symptoms developing at tips and margins of leaves. Necrosis is characteristic, with older to middle-aged leaves affected first in many species. Chlorosis may be the first symptom. In young leaves, chlorosis is followed by a water-soaked condition and later by necrosis and desiccation of tissues. Tan, light brown, or dark brown colorations often develop.

TABLE 5.7   Susceptibility of Various Plants to Atmospheric Fluoride

Species	Common Name
Very high	
Asparagus densiflorus 'Sprengeri'	Sprenger asparagus
Chlorophytum comosum 'Variegatum'	Spider plant
Euphorbia pulcherrima	Poinsettia
High	
Begonia rex-cultorum	Begonia
Chamaedorea elegans	Parlor palm
Coffea arabica	Coffee
Cordyline terminalis 'Baby Doll'	Ti
Dracaena deremensis 'Warneckii'	Striped dracaena
Lilium longiflorum	Easter lily
Philodendron bipennifolium 'Panduriforme'	Fiddle-leaf philodendron
Pteris cretica 'Albo-lineata'	White-line Cretian brake
P. cretica 'Mayii'	May Cretian brake
P. ensiformis 'Evergemensis'	Evergemen table fern
Moderate	
Chrysalidocarpus lutescens	Areca palm
Dizygotheca elegantissima	False aralia
Dracaena deremensis 'Janet Craig'	Janet Craig dracaena
Howea forsterana	Kentia palm
Maranta leuconeura erythroneura	Prayer plant
Philodendron scandens oxycardium	Heart-leaf philodendron
P. selloum	Lacy-tree philodendron
Phoenix roebelenii	Phoenix dwarf palm
Rhoeo discolor	Moses-in-a-boat

TABLE 5.7   (Continued)

Species	Common Name
Low	
Aglaonema commutatum 'Silver Queen'	Silver evergreen
Aphelandra squarrosa 'Dania'	Zebra plant
Araucaria heterophylla	Norfolk Island pine
Ardisia crenulata	Coral berry
Brassaia actinophylla	Schefflera
Dieffenbachia amoena	Dumbcane
Dracaena fragrans 'Massangeana'	Corn plant
Hoya carnosa	Wax plant
Nephrolepis exaltata 'Fluffy Ruffles'	Fluffy ruffles fern
Philodendron domesticum 'Red Emerald'	Red emerald philodendron
Pilea cadierei	Aluminum plant
Podocarpus macrophyllus	Podocarpus, southern yew
Sansevieria trifasciata 'Hahnii'	Snake plant
Sinningia speciosa	Gloxinia
Syngonium podophyllum 'Green Gold'	Nephthytis
Zebrina pendula	Wandering Jew
Very low	
Chrysanthemum morifolium	Chrysanthemum
Codiaeum variegatum	Croton
Epipremnum aureum	Pothos
Gynura procumbens	Purple passion vine
Nephrolepis exaltata 'Bostoniensis'	Boston fern
Peperomia caperata	Crinkled peperomia
P. obtusifolia 'Variegata'	Variegated peperomia
Plectranthus australis	Swedish ivy

Source: Adapted from S. S. Woltz and W. E. Waters, HortScience 13(4): 430–432; 13(5): 585–586, 1978.

When foliage plants are installed in conjunction with an enclosed swimming pool, chlorine injury may develop because of vaporization of the chlorine. Common symptoms of chlorine toxicity are necrosis and bleaching of leaves.

In addition to maintaining quality water in pools, sodium hypochlorite (NaOCl), bleach, is used for general cleaning, alone or in combination with other chemicals. Poole and Henley measured the effects of short-term exposure of plants to both NaOCl solution fumes and drenches. The species tested were Dizygotheca elegantissima, Dracaena deremensis 'Janet Craig', Ficus benjamina, Ficus elastica 'Robusta', Homalomena 'Emerald Gem', Philodendron scandens 'oxycardium', and Polyscias fruiticosa. The fumes did not visibly injure any of the species tested. When the growing medium was drenched with bleach solution, leaves of recent and new growth of Homalomena 'Emerald Gem' and Ming aralia (Polyscias fruiticosa) showed necrosis significant enough to reduce plant quality from excellent to poor. The other species were not affected. 'Emerald Gem' *Homalomena* and Ming aralia would be inappropriate plants for plantscapes near pools, spas, or other water accents where they might be splashed with chlorinated or chemically treated water.

Ozone in the building interior may also cause foliage injury to susceptible plants. Copy machines and laser printers produce significant amounts of ozone. In addition, ozone may be drawn into buildings through the fresh-air intakes of ventilation systems. Skelly, reported by Clein (1994), has shown that *Dracaena deremensis* 'Warneckii' and *D. deremensis* 'Janet Craig' have the potential to reduce ozone levels indoors.

Recent investigations by Poole and Conover (1991) show that mercury (phenylmercuric acetate), used as a mildewcide in many paints, will cause severe leaf drop in susceptible species. Other symptoms include chlorosis of lower leaves followed by necrosis; small, thin, pale, soft and viscous new green leaves; thin branches that wilt and die; and stunted and deformed growth. *Ficus* and *Dieffenbachia* species are most susceptible with increased exposure time increasing injury (Figure 5.6). Since paints containing phenylmercuric acetate will probably emit enough mercury for a year after painting to seriously defoliate *Ficus* species, interiorscapers should suspect this pollutant when severe leaf drop occurs in a newly painted area. Mercury levels causing plant injury were well below established government limits for human environments, and health problems are not a concern.

Most interior plantscapers maintain a facility for holding plants for replacements or new installations. They vary from greenhouses to basements, garages, and specially designed "warehouses." Careful consideration must be given to the heating system and location of heaters.

In small areas, space heaters, especially those burning kerosene, may be an economical way of maintaining proper temperatures. All space heaters require adequate ventilation.

Unvented kerosene heaters require the use of clear white or 1-K grade kerosene. This fuel is 99.5% clean burning and has a maximum sulfur content of 0.04 weight percent. The 1-K grade is costly. The second grade is 2-K which has a maximum sulfur content of 0.3 weight percent. Without proper ventilation, burning 2-K grade kerosene increases carbon monoxide and sulfur dioxide in the atmosphere. Carbon monoxide can be life-threatening because it robs the blood of oxygen and prevents removal of carbon dioxide.

Sulfur dioxide (SO_2), although not life-threatening, can injure plants. Exposure to low concentrations of SO_2 will cause bleaching and burning of leaves of susceptible species. Interveinal tissue is most affected with the injury usually most prominent toward the petiole. Fully expanded leaves are most susceptible.

Finally, if a heater is unventilated, and there is little air seepage into the area, oxygen may be reduced to a level where combustion cannot occur. The flame goes out and there is no heat. As the temperature drops, foliage plants may be injured, or killed. Other gases in the atmosphere are not apt to be a problem unless their concentrations are very high.

(a)

(b)

Figure 5.6 Leaf drop caused by off-gassing of mercury from paint. (a) *Ficus benjamina* 7, 21, 35, and 49 days after exposure (right to left); (b) *Dieffenbachia* 'Camille' 7, 21, 35, and 49 days after exposure (left to right).

Unless it is filtered, the air in our building interiors will contain dust and dirt of various types. These particles will settle on plants and mar their beauty. Dirt may also clog stomata, pores in the leaf through which gas exchange occurs, and must be removed periodically. Regular use of a simple feather duster will clean plants. If necessary, washing the leaves with mild soap and water, or a solution of nonfat milk, will remove dirt. Commercial plant shines should be used sparingly, as such product may build up on the leaf surface, plugging the stomata. In addition, they may be sticky and attract more dirt. Since the leaves of many tropical foliage plants have a waxy layer of cutin on the surface which has a natural shine if kept clean, use of additional waxes is usually not necessary.

SUMMARY

The atmosphere consists of the gaseous envelope surrounding the plant and the gases in the growing medium and within plant tissue. Oxygen and carbon dioxide are essential for respiration and photosynthesis, respectively, and are present in adequate amounts in the indoor atmosphere. Oxygen deficiency in soil, due to overwatering, may kill or injure the roots. There is no danger of plants depleting all of the oxygen in a room.

Plants will adapt in time to the low relative humidity of building interiors. Methods used to increase the water-vapor content of the air include installation of humidifiers, grouping of plants, placing plants above trays of water or in high-humidity areas, the use of pools and fountains, and closed growing environments.

Sick building syndrome is a human health problem generally related to air pollution indoors. Plants reduce indoor air pollutants, especially formaldehyde, benzene, trichloroethylene, xylene, and ammonia. Other pollutants, especially ethylene, may be a problem at times. Dust and dirt which collect on the plants mar their beauty and interfere with the exchange of gases, and should be removed.

REFERENCES

CLEIN, M.: "Indoor Ozone Results Released Following Year-Long Study," *Interiorscape* 13(2): 22, 1994.

_____: "Move Over Spock, Dr. Bill is Back," *Interiorscape* 12(4): 60–63, 1993.

CONOVER, C. A. and R. T. POOLE: "Response of Foliage Plants to Commercial Interior Paints," *Foliage Digest* 10(1): 4–5, 1987.

FISHER, C. C., and R. T. FOX: *The Selection, Care, and Use of Plants in the Home,* Information Bulletin 117, New York State College of Agriculture and Life Sciences, Cornell University, Ithaca, N.Y., 1977.

GAINES, R. L.: *Interior Plantscaping,* Architectural Record Books, New York, 1977.

GRAF, A. B.: *Exotic Plant Manual,* Roehrs Company, East Rutherford, N.J., 1970.

Indoor Air Facts No. 4, Sick Buildings, United States Environmental Protection Agency, Washington, D.C., 1988.

MAROUSKY, F.: "Effects of Ethylene in Combination with Light, Temperature, and Carbon Dioxide on Leaf Abscission in *Fittonia verschaffeltii* (Lem.) Coem. var. *argyroneura* (Coem.) Nichols," *Proceedings of the Florida State Horticultural Society* 92: 320–321, 1979.

_____ " The Shipping Environment," Lecture, 1978 National Tropical Foliage Short Course, Orlando, Fla.

_____, and B. K. HARBAUGH: "Interactions of Ethylene, Temperature, Light, and CO_2 on Leaf and Stipule Abscission and Chlorosis in *Philodendron scandens* subsp. *oxycardium,*" *Journal of the American Society for Horticultural Science* 104(6): 876–880, 1979.

POOLE, R. T.: "Ethylene-Sensitivity and Sources," *Foliage Digest* 9(11): 8, 1986.

POOLE, R. T., and C. A. CONOVER: "In the Dark about Foliage Plants?" *Florida Foliage Grower* 14(11): 1, 1977.

_____: "Mercury Toxicity to *Ficus* spp.," *Foliage Digest* 17(7): 7, 1991

_____ and R. W. HENLEY: *Reaction of Seven Foliage Plants to Sodium Hypochlorite Fumes and Drenches,* University of Florida, Central Florida Research and Education Center-Apopka, CFREC-Apopka Research Report RH-92-95.

REBER, P. N.: "Are Kerosene Heaters Safe in Greenhouses?" *Green Scene* 11(3): 30, 1983.

WOLTZ, S. S., and W. E. WATERS: "Airborne Fluoride Effects on Some Flowering and Landscape Plants," *HortScience* 13(4): 430–432, 1978.

_____: "Airborne Fluoride Effects on Some Foliage Plants," *HortScience* 13(5): 585–586, 1978.

WOLVERTON, B. C., A. JOHNSON, and K. BOUNDS: *Interior Landscape Plants for Indoor Air Pollution Abatement,* National Aeronautics and Space Administration, John C. Stennis Space Center, Stennis Space Center, Miss., 1988.

_____ and J. WOLVERTON: "Interior Plants and Their Role in Indoor Air Quality," *Interiorscape* 11(4): 16–21, 1992.

Planters

Selection of planters for interior landscapes requires careful consideration, as they must not only be functional, but decorative as well. The container provides the growing area for the plant's roots, and holds the growing medium. It may also hold and hide the growing pot.

Just about any container can be used or adapted to grow plants. There are various types, ranging from very expensive to those that may be "free." Planters are available in sizes for individual plants or groupings. They may be portable or built-in, with or without drainage.

CHOOSING A PLANTER

In selecting planters for the interior plantscape, consider both practical and aesthetic qualities. In commercial installations, the architect or interior designer will probably stipulate the size, type, color, and texture of the containers to be used, and plants will be selected to complement them. For ease of replacement, it is generally better to keep the plant in its plastic or metal production container and place it inside a decorative pot or box, rather than plant it directly.

The following should be considered when choosing a planter:

1. How well it suits the needs of the plant.
2. How well it suits the needs of the individual and the environment.
3. Cost and availability.
4. Strength and durability.

5. Weight.

6. Drainage.

Suiting the Needs of the Plant.   To a horticulturist, the most important consideration is how well the container suits the needs of the plant. If used for direct planting, it must be sufficiently large so as not to restrict the root system. It must accommodate the growing pot when used as a jardiniere or in double-potting. All planters should be in proportion to the height and width of the plant. In general, tall plants look better in a tall container, while broad, shrublike specimens should be placed in a lower, wider receptacle (Figure 6.1). Shallow pots are used to proportion the planting and deemphasize height. Thus, selection of the planter and the plant should be made at the same time.

Before specifying or ordering decorative containers, one should know the actual inside diameter and height. A lip may reduce the width and the grow pot will not fit: a larger size will be needed. If the bottom is fluted or ribbed, inside height may be reduced causing the grow pot to protrude above the top of the jardiniere, creating an unsatisfactory appearance.

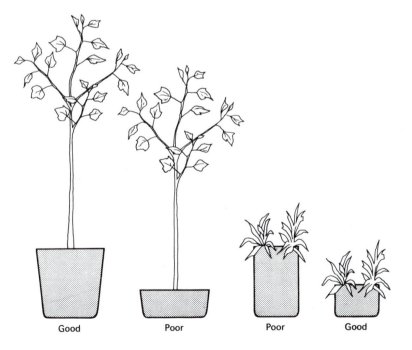

Figure 6.1   Planters must be in proportion to the height and width of the plant.

Suiting the Needs of the Individual and the Environment. The decorative value of the planter must also be considered. Its style, color, texture, and proportions will be determined by the plant chosen and where it is to be placed. The decor of the area—room, office, corridor—will influence the choice, as may the client's preference. Containers made from a variety of materials are available in sizes, colors, and styles to satisfy most decorating needs. Custom-made planters are also a possibility. Appendix A lists some sources of decorative planters for interior plantscape specialists.

If the planters are to be used as part of a grouping, it is best that they match. A plant displayed by itself may permit use of a planter that is "one of a kind."

Cost and Availability. When purchasing containers, cost will usually be a consideration. While price may not be important when buying one or two planters, planters represent a major investment in large installations. The planters must be easily obtainable. For the home gardener, the selection will probably be made from those planters on display at the retail shop. For the professional working with a client, the planters must be selected and ordered, and in location at the site on the date the plants are to be installed.

To solve potential availability problems, many firms select a container style that is generally appealing and maintain an inventory. Although such practice uses valuable storage space and ties up cash, instant access to planters may justify the added costs.

Strength and Durability. The planters should be durable—sufficiently strong to hold the plant and the growing medium and to withstand normal wear and tear without cracking or chipping. They should be colorfast and easy to clean.

Weight. Weight is also important. Are the containers heavy in their own right? Remember: when filled with a plant and its wet growing medium, considerable weight will be added. Will the building support the weight, or must lighter materials be used? Will container be moved? If so, what provisions must be made to move it?

Drainage. Planters are available both with and without drainage. If holes are provided in the base, some provision must be made to catch the excess water. Not to do so will result in damage to floors, carpets, or furniture. Setting the containers on nonporous trays or saucers will solve the problem. Where no drainage is provided, the planters are usually used as decorative containers or jardinieres, holding plants in pots with drainage. Applying too much water may be a problem, so the inner pot should be raised slightly to provide a reservoir for excess water below it.

Planting directly inside a container without drainage is possible but is not recommended unless provision is made for the excess water (Figure 6.2). Containers without drainage cannot be leached, and fertilizer practices may have to be modified to prevent buildup of salts. All planters should be kept clean.

Raising decorative planters slightly above floors, tops of desks, tables, files, and other surfaces will eliminate possible condensation underneath and prevent

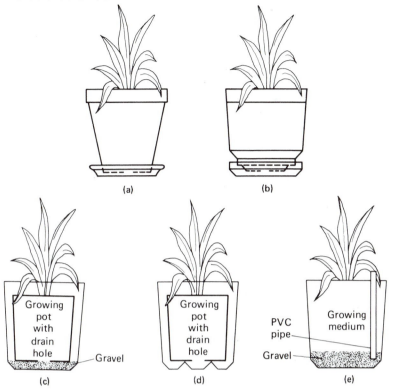

Figure 6.2 Providing for excess water. Planters (a) and (b) have saucers; (c), (d), and (e) are watertight and collect water in the bottom.

moisture damage to the surface. Acrylic disks, protective mats, or legs of some sort are satisfactory for this purpose.

MATERIALS FOR CONTAINERS

Planters for interior gardens may be constructed of various materials or combinations of materials, including wood, clay, plastic, fiberglass, ceramic, concrete, stone, glass, and metals, including brass, aluminum, and iron.

Wood

Wooden planters consist of tubs, boxes, or barrels made from cedar, cypress, redwood, black cherry, black walnut, black locust, white oak, and exterior plywood (Figure 6.3). Redwood is the most popular. Because they are organic, wooden containers are subject to decay and may be flammable.

(a)

(b)

Figure 6.3   Wooden planters; *Ficus retusa* 'Nitida'.

When using home-built wooden planters indoors, never treat the wood with creosote or pentachlorophenol preservatives. To do so is illegal unless the wood has been sealed with two coats of an appropriate sealant such as urethane, epoxy, or shellac after applying the preservative. Both are toxic and may injure or kill plants. Creosote may leach into the soil. Fumes from either may bleach foliage, with marginal and tip yellowing often the first symptoms. Foliage that contacts treated wood is especially susceptible.

Pressure-treated wood has been treated with either of two inorganic arsenical compounds, chromated copper arsenate (CCA) or ammoniated copper arsenate (ACA). They bind tightly to the wood and have a very low tendency to leach into the growing medium. Arsenical-treated lumber is not injurious to plants. Copper or zinc naphthanate (Cuprinol) treated wood may also be used safely.

Softwoods treated with brushed-on preservative are more durable than the same wood untreated, but not as durable as hardwoods that have been pressure treated. The life of any wood can be extended by painting, varnishing, or staining. To prevent dry rot, do not paint the base or nonexposed ends of boards. The bottoms of wooden containers should be raised off the floor to permit air circulation and reduce rotting.

To further maximize the life of a wooden planter, the inside should be lined with a plastic or rustproof sheet-metal insert. Purchase wooden planters with inserts if possible. A double layer of 4- or 6-mil polyethylene plastic lining the inside and carefully stapled below the rim of the planter should prove satisfactory.

A watertight galvanized steel insert soldered at the corners makes an excellent liner. A sheet-metal contractor can construct such inserts to specifications. Aluminum inserts are easily made and quite suitable. If the area of the planter is large, more than one insert may be needed. To make an aluminum insert, purchase a 3- by 3-ft sheet of aluminum from a hardware store or building supply center. Determine the size of insert needed and map it out on the aluminum as illustrated in Figure 6.4. Draw a rectangle or square the size of the base (ABCD). Determine the depth. Draw a second rectangle or square around the first at a distance equal to the depth of the insert (EFGH). Pencil in the corners by extending all lines of the inner rectangle. Cut along the outermost lines. Clamp a strip of wood along the line joining side and base on two perpendicular sides, for example AB and BC. Carefully bend the sides perpendicular to the base. As you begin folding the sides upward, crimp the corner so that it points away from the tray at 45°. Do the same with the other sides and corners. The start of four "breadpan" folds is now apparent. To complete the corners, flatten the folds against the sides of the insert using a block of wood to support the corner from the inside, while hammering from the outside. File any exposed, rough edges.

To prevent corrosion of either galvanized steel or aluminum from salts in the drainage water, paint the inside of the insert with asphalt before installing in the planter.

Clay

Unless they have been made waterproof, planters made from clay or terra-cotta (Figure 6.5) are porous and permit evaporation of water. The growing medium dries quickly, necessitating more frequent watering. As much as 50% of applied water will evaporate directly through the sides of clay pots. However, if a poor growing medium

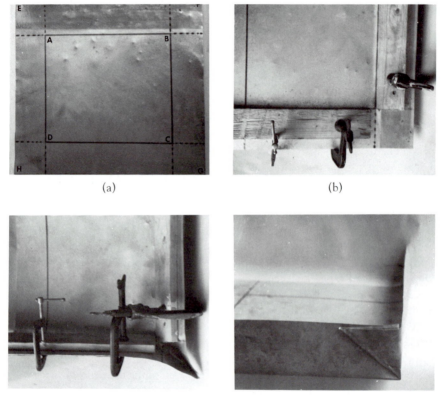

(a)   (b)

(c)   (d)

Figure 6.4   Construction of a metal insert for a wooden planter: (a) insert
measured on metal; (b) clamps and bending strips in place; (c) sides bent
upward; (d) corner completed. Follow the same procedure for the other three
corners.

is used, porous clay pots foster better soil aeration and reduce the potential of plant
injury due to watering too often.

Clay containers are usually an earthy brown color. They are available in various sizes and may have modifications in shape ranging from the standard "pot," to
modern, to reproductions of classic designs. Ordinary pots, either clay or plastic, are
available as standard pots, azalea pots, or pans. Standard pots are as high as they are
wide, an 8-in. standard pot being 8 in. high and 8 in. across. Azalea pots are three-
fourths as high as wide, with an 8-in. azalea pot being 6 in. high and 8 in. wide at the
top. Pans are one-half as high as they are wide. Because they have broader bases, the
more stable azalea pots or pans are preferred for use indoors.

Clay planters are heavy, especially after watering, and break more easily than
pots made of other materials. Dry clay planters absorb considerable amounts of

Figure 6.5 Terra-cotta planters: (left) *Chamaedorea erumpens* underplanted with *Philodendron* 'Emerald Queen'; (right) *Fatsia japonica*.

water, so they should be soaked thoroughly before using. Clay is harder to keep clean than are other types of containers because algae will grow on the constantly moist surface. Also, as water evaporates, salts are deposited on the surface of the clay, particularly near the rim, forming a white crust and permanently discoloring the pot. Clay planters may quickly lose their attractiveness.

Clay saucers are porous and allow water seepage, which may mar furniture or cause rotting of carpets. Raise clay saucers from the surface to eliminate this problem.

Nonporous Planters

Various planters made from nonporous materials are on the market. Because their sidewalls are water-impervious, evaporation is reduced and they require less frequent watering. They may be more easily overwatered. The air above the planter is not as humid as with clay, again because of reduced evaporation.

Plastic.   Numerous types of plastics, molded into nonporous planters, are available in a wide range of shapes, sizes, colors, and designs (Figure 6.6). Plastics are light in weight and less expensive than clay, and can be watertight. They are colorfast and may be painted to complement any interior design. Most types are strong and shatter-resistant, although some become brittle and crack easily with time, perhaps as a result of exposure to ultraviolet light or high temperatures.

Plastic planters are easily cleaned because algae do not grow on them, nor do salts accumulate.

Figure 6.6   Plastic planter with *Yucca elephantipes*.

Clear and white plastic pots are transparent and translucent, respectively, and may permit algae to grow on the surface of the medium adjacent to the pot wall. The presence of algae can impede drainage and soil aeration. In addition, light shining through the pot wall has been reported to inhibit the growth of roots at the surface of the medium. Thus, the root system will not penetrate the entire substrate mass and will have to absorb water and nutrients from a smaller volume of medium. Roots of *Chamaedorea* and *Alglaonema* are especially responsive to light. In a more recent study, Boodley (1978) grew 10 different foliage plants in seven different glass or ceramic containers. There was no significant difference in root and top growth among the treatments, nor did roots grow away from the light. These results notwithstanding, greatest satisfaction is usually achieved by placing white or clear pots inside an opaque jardiniere.

Fiberglass.   Fiberglass has glass fibers embedded in a plastic resin, producing a strong planter with most of the virtues of plastic (Figure 6.7). Fiberglass planters are molded without seams and are leakproof and colorfast. They are very attractive and available in virtually any shape and color, with sizes up to 8 ft. Containers with a molded lip that curves into the planter look better than those with straight sides. Fiberglass may be used to copy the appearance of other materials, such as metal or stone, thereby allowing the use of these elements in interior landscapes without the

(a)

(b)   (c)

Figure 6.7 Fiberglass containers are extremely decorative and come in a wide range of sizes and colors: (a) *Howea forsterana;* (b) *Dracaena deremensis* 'Janet Craig'; (c) *Ficus retusa* 'Nitida'. (Courtesy of Pouliot Designs Corp.)

weight of the real product. Although relatively inexpensive, fiberglass planters are more costly than molded plastic containers of the same size.

Glazed Clay. Glazed clay containers are similar to clay, but evaporate less water because the surface is sealed in the glazing process.

Ceramic. Many attractive planters, available in a wide variety of sizes and shapes and infinite colors and textures, are made of ceramic (Figure 6.8). They are heavier than plastic and, unlike clay, evaporate little water. Ceramics are subject to breakage.

Concrete. Portable concrete containers are not well suited for use indoors, as they are usually very heavy. If they are used, vermiculite or perlite should be added to the mix to reduce the weight.

In commercial installations, especially malls, fixed planters of concrete or masonry are frequently used (Figure 6.9). If possible, drains should be built in for the

(a)

(c)

(b)

Figure 6.8   Ceramic planters: (a) *Dracaena marginata*; (b) *Chamaedorea erumpens*; (c) *Polyscias guilfoylei*. (Courtesy of Everett Conklin.)

(a)

(b)

Figure 6.9 Concrete planters: (a) above ground with tiled walls, *Codiaeum variegatum* 'pictum'; (b) inground, *Brassaia actinophylla* with an underplanting of *Dracaena deremensis* 'Warneckii'.

removal of excess water. If drains are not installed, provision for excess water must be made when planting.

Stone.   Planters made of stone are usually installed as a permanent element in the design of the interior plantscape. They are frequently used in commercial locations such as malls and lobbies, where provisions have been made for the weight.

Glass.   A wide variety of glass containers, mostly for home use, are available, ranging from aquaria and fishbowls, to brandy snifters, jars, and bottles of various types. Glass is easily broken. Roots may not grow against the glass surface, so that the root system may be further constricted in an already small volume of growing medium. Glass containers are frequently used for hydroponic culture.

Metal.   Metal is rarely used for direct planting because of potential toxicity. When using a brass container, the inside of which is untreated, a lining should be provided. The copper from the brass is toxic to plants. Aluminum planters, shown in Figure 6.10, make excellent jardinieres.

(b)

(a)

Figure 6.10   Aluminum planters: *Dracaena fragrans* 'massangeana' as (a) a multicaned specimen and (b) a desktop plant.

Decorative Aspects

Baskets made from reed, grass, willow, and other materials are available at various prices in a wide range of sizes, shapes, colors, and surface textures (Figure 6.11). They make attractive covers for growing pots. Being organic, baskets will tend to rot in time. Rotting may be slowed by using watertight inner pots, placing saucers inside the basket under growing pots with drainage to catch excess water, or lining the base of the basket with plastic.

The basket may also be waterproofed by lining the bottom and the lower sides with a coating formed of several layers of polyester resin and newspaper strips.

Decorative sleeves covered with burlap, reed, and other materials are also very effective in hiding an unattractive growing pot.

Attractive, functional planters for interior plantscaping projects, large and small, are readily available. Plastic, fiberglass, and wood are most frequently used both commercially and at home. At home, attractive, unique planters may be made from many of the "odds and ends" that clutter the attic or basement, or which may be found at flea markets or antique shops. Be on the lookout for items that may make unusual planters. Do not discount such everyday things as buckets, wastebaskets, pots and pans, cans, lettuce baskets, barrels, and washtubs. How about TV dinner trays, milk cartons, margarine containers, and "Big Mac" boxes? Building materials such as flue liners, building blocks, drain tiles, and sewer pipes may also have a place. Coal scuttles, milk cans, and butter churns all make excellent planters. The list is endless; all one need do is use a little imagination.

(a)

(b)

Figure 6.11 Baskets are attractive covers for growing pots: (a) *Ficus elastica* 'Decora'; (b) *Asparagus densiflorus* 'Sprengeri'.

HANGING PLANTERS

Foliage plants for the interior garden are no longer relegated to the floor, desk and tabletop, or windowsill. They are hung in great profusion from walls, ceilings, and window frames (Figure 6.12).

Several types of hanging planters are available, including those with saucers, without drainage, and wire baskets.

The first two types are suspended from hooks or brackets and have a reservoir for collecting excess water, the first in the saucer, the second in the base of the planter. Wire baskets are open frames lined with sphagnum moss. The moss serves to contain the growing medium and the plant roots. They are extremely difficult to water, and there is no place for excess water to accumulate, so that water drips freely on carpets, furniture, or whatever is beneath the basket. Wire baskets have no place in the indoor garden.

Hanging planters are heavy, with 1 ft^3 of wet soil weighing up to 90 lb. Weight may be reduced by using the lighter peat-lite or bark-amended growing media, and plastic or fiberglass containers. In any event, the hooks and brackets, as well as the hanger cords and building, must be strong enough to support the weight of the entire planting.

Hooks should be of sufficient size and screwed directly into a stud in the wall or joist in the ceiling. Where this is not possible, use some type of anchor. In ceilings, drill a hole and install either a toggle bolt or a Molly-type screw to which the hook may be attached. In hollow walls, use hollow-wall anchors to which the brackets may be attached. Swivel hooks are available which facilitate turning plants.

In brick, concrete, or cinder-block walls, vinyl or metal anchors may be used. Drill a hole just large enough for the anchor, then insert and tighten the screw. The anchor will expand and hold the bracket firmly in place. Plant tracks are also suitable provided that they are anchored directly to the joists or attached with hollow-wall anchors of the proper size. In commercial plantings, the selecting and installation of hooks and anchors should be left to the engineer and contractor.

Caution should be used in hanging planters from the metal grid of suspended ceilings. There is a minimum amount of support holding up such framework. Additional bracing may be needed before hanging a plant to prevent bending of the frame from the weight.

Hangers should be strong, and include leather thongs, macramé cord, hemp, wire, chain, and metal hooks. Hangers that permit one to lower the plant for watering and grooming are recommended for locations that are relatively inaccessible.

HOME-BUILT PLANTERS

Suggestions and plans for planters you build yourself are available in many publications, including books, craft manuals, and gardening and hobby periodicals, or you may design your own. They are usually customized for a specific location and may be

(a)

Figure 6.12　Hanging planters:
(a) aluminum, *Epipremnum aureum*;
(b) ceramic, *Nephrolepis exaltata*
'Florida Ruffles'; (c) plastic,
Chlorophytum comosum 'vittatum'.

(b)

(c)

simple or elaborate, depending upon the decor. Various materials, including wood, plastic, stone, and brick, may be used in the construction.

Weight must be a consideration when constructing large planters. The floors in most homes are designed to support a live load of 40 lb/ft^2. A cubic foot of wet soil may weigh up to 90 lb plus the weight of the container and plant, thereby creating a heavy load.

In constructing planters, waterproof materials should be used or the surface protected with several coats of polyurethane varnish. The inside should be lined with polyethylene, or a metal insert installed. Raising floor planters slightly will permit air circulation and prevent the accumulation of moisture.

SUBIRRIGATION (SELF-WATERING) PLANTERS

Subirrigation planters are intended to minimize moisture stress in interior plantings. They were first used in Europe, Australia, Africa, and the Middle East over 30 years ago and are now used in significant numbers in interiorscapes in the United States.

Made from plastic or fiberglass, self-watering planters are available in a variety of sizes, styles, and colors (Figure 6.13). They usually consist of a double-walled pot with a water reservoir that is filled every 2 to 4 or 6 weeks, thus reducing watering frequency. When properly managed, plants experience reduced water stress due to over- or underwatering, resulting in long-term maintenance of good-quality plants with fewer replacements. Compaction of the growing medium is reduced because there is no overhead watering; thus aeration is better. Nutrients may be incorporated into the water.

There are several other advantages associated with these types of planters. Maintenance time is saved, thereby reducing costs and enabling the technician to service more accounts. There is also more time for other chores, cleaning the plants, for example. Use of subirrigation planters can ease the watering problems associated with employee turnover or when substitutes fill in during absence of the regular technician.

Self-watering containers are excellent where the water source is inconvenient or for plants located in inaccessible areas such as conference rooms, executive offices, residences, and hard-to-reach areas, including ledges, balconies, and hanging planters. They are used by many interior plantscape firms for seasonal flowering plants such as pot mums, which require frequent watering, and may make servicing sites with numerous small plants more profitable.

Because there are fewer visits to the site, there is less interference by the plantscape technician with the normal routine of the facility and fewer distractions to the workers. Commercial firms also have better security control in their buildings because fewer people enter and leave the facility.

Subirrigation planters are not without their disadvantages. Cost savings may not be as great as first supposed. It takes time to fill the water reservoir. Does one filling a month take less time than watering once a week? It may not. Long intervals

Figure 6.13 Subirrigation (self-watering) planter. A double-walled pot forms the water reservoir, which is filled periodically by removing the stopper.

between visits to the site reduce the opportunity for the technician to inspect the plants for pests, diseases, and other problems. Left undetected, corrective measures may be delayed.

These planters are not for every plant. Cacti and succulents are easily overwatered. Also, they may not be compatible with every decor, and some may be more expensive than other types of containers.

Because plants are usually direct planted in self-watering containers, the growing medium is critical. If not loose with numerous large pores, overwatering may result. Leaching of soluble salts that accumulate near the surface may be difficult with some self-watering systems.

There are several types of self-watering planters available commercially. All are capillary systems permitting water to rise in the small pores of the growing medium in a manner similar to absorption by a paper towel. Most have a separate water reservoir and use wicks of fabric or the growing medium to move the water. A few use traditional hydroponics with the plant's root system partially immersed in water. The operation of subirrigation planters is described in Chapter 9.

PRODUCTION CONTAINERS

Production containers or grow pots are the plastic, fiberglass, or metal planters in which the plants are grown in the greenhouse or nursery. They are unattractive and

TABLE 6.1   Production Container Sizes

Pot Size (in.)	Trade Designation	OAW × OAH Actual Dimensions (in.)
6	1 gal std.	6½ × 6
6-in. Azalea	6-in. tub	6 × 4¾
8	2 gal	8 × 8
8-in. Azalea	8-in. tub	7½ × 5¾
9	9-in. pot	8¼ × 8
10	3 gal	10 × 9
10-in. Azalea	10-in. tub	10 × 7½
12	4 gal	11½ × 11
14	7 gal	14 × 12
17	10 gal	17 × 14¾
22	20 gal	21½ × 17
28	45 gal	28 × 17¾
30	50 gal	29 × 19½
32	65 gal	31½ × 20
38	95 gal	37½ × 25
42	100 gal	41 × 24½
48	150 gal	49 × 25½
50	200 gal	50 × 27
60	300 gal	61 × 32

not meant to be exposed in the interior plantscape. Grow pots may be hidden by inserting them into decorative planters or by double-potting. They are removed when direct planting. Trade designations for production containers are shown in Table 6.1.

SUMMARY

Planters are both decorative and functional. They should suit the needs of the plant as well as the needs of the individual and the environment. Cost and availability, strength and durability, weight, and drainage should also be considered. Suitable portable or built-in planters may be made of wood, clay, plastic, fiberglass, ceramic, concrete, glass, and metal. Just about any container can be adapted to growing a plant. Filled planters are heavy and weight must be considered, particularly with large plantings and hanging plants. Sufficient support must be provided. Custom planters are easily built for home and commercial locations. Subirrigation planters reduce the frequency of watering. Properly managed, they minimize moisture stress in interior plantings, and may reduce maintenance time and cost. Grow pots are removed when planting, or hidden in decorative planters.

REFERENCES

BOODLEY, J. W.: "Growth of Selected Plants in Opaque, Transparent and Translucent Containers" (abstract), *HortScience* 13(3): 350, 1978.

CONOVER, C. A. and R. T. POOLE: "Foliage Plant Responses to Translucence of the Growing Container," *HortScience* 14(5): 616–617, 1979.

COPLEY, K.: "A Guide to Self-Watering Planters," *Grounds Maintenance* 16(7): S-8 to S-13, 1981.

Editors of Sunset Books and Sunset Magazine: *Plant Containers You Can Make*, Lane Publishing Co., Menlo Park, Calif., 1976.

GRIESS, K.: "Self-Watering Pots Can Save Money," *Southern Florist and Nurseryman* 95(23): 24–25, 1982

"Growing Gracefully," *Popular Mechanics* 143(4): 120–121, 1975.

HAMMER, N.: "Selecting the Right Container," *Interiorscape* 12(3): 6, 72–73, 76, 1993.

JOHNS, L.: *Plants in Tubs, Pots, Boxes and Baskets*, Van Nostrand Reinhold Company, New York, 1974.

NORMAN, C.: "Wood Preservatives," *Green Scene* 15(2): 31–33, 1986.

POULIOT, T.: "Revolution in Planters," *Landscape and Turf* 26(6): 29, 1981.

Professional Guide to Green Plants, Florists' Transworld Delivery Association, Southfield, Mich., 1976.

SCHARFF, R.: *The Book of Planters*, M. Barrows and Company, Inc., New York, 1960.

SCHULDENFREI, P.: "Working from the Bottom Up," *Florists' Review* 173(4483): 29–31, 1983.

STAT, R.: "A Pebble Tray," *House Plants and Porch Gardens* 3(2): 72–74, 1978.

WHITCOMB, C. E.: "Drainage Factors in Plant Containers," *Florists' Review* 156(4051): 23, 56–57, 1975.

WHITE, J. W., and J. W. MASTALERZ: "Container Gardening Offers Something for Everyone," *Landscape for Living*, The Yearbook of Agriculture 1972, House Document No. 229, Washington, D.C.

WHITE, M. G.: *Pots and Pot Gardens*, Abelard-Schuman Ltd., London, 1969.

WOODHAM, T.: "Keys to Selecting the Right Container," *American Nurseryman* 156(10): 75–76ff., 1982.

The Growing Medium

Plants in the interior plantscape require a constant supply of water and minerals for normal growth. Although limited quantities may enter the plant by way of the leaves, almost all of the water and minerals are absorbed by the roots from the substrate in which they are growing. A suitable growing medium, then, must not only provide sufficient amounts of water and essential elements for the plant's needs, but it must also provide an environment suitable for the growth and functioning of the root system.

FUNCTIONS OF THE GROWING MEDIUM

The primary function of the growing medium is to provide water and essential minerals to the plant. Water is supplied by periodic irrigations and, following drainage of excess, absorption of the stored water by the plant over time. Essential minerals are derived from the parent material of the medium, decay of organic matter, or applied as fertilizer.

The roots of the plant permeate the growing medium. Roots are living organs and require oxygen for respiration. Since plants, unlike higher animals, do not translocate oxygen internally, the substrate is the source of all the oxygen used by roots.

In addition, the medium anchors the plant and enables the stem to grow upright without toppling. If suitable support and oxygen are provided, the growing medium could be water.

CHARACTERISTICS OF A GOOD GROWING MEDIUM

A satisfactory medium for interior plantings will be porous and well drained, and yet retain water sufficient for the plant's requirements. As a general rule, horticulturists say that a soil-based substrate should consist of 50% solid particles and 50% pore space. Following irrigation and drainage, 50% of the pores should be filled with water and 50% with air. A soilless medium retains 20 to 60% of its volume as water.

The growing medium should be low in soluble salts (see Chapter 8) but have sufficient cation-exchange capacity (C.E.C.) to retain and supply fertilizer elements. A C.E.C. range of 2 to 40 milliequivalents per 100 grams (g) of dry medium will retain adequate amounts of fertilizer without injuring plant roots. Media with high C.E.C. values are strongly buffered; their pH is not easily changed. The substrate should be standardized and easily duplicated, so that watering, fertilizing, and other maintenance operations may be routine. Different media require different practices, thus increasing the time and cost of maintaining the plants.

The weight or bulk density determines the ease of handling and the ability of the medium to keep the plant upright. A bulk density ranging from 0.15 to 0.75 g/cm^3 is desirable. Volume wet or dry should remain fairly constant. Shrinkage should not exceed 10%.

In addition, a suitable medium should be uniform; free of diseases, pests, weed seed, and harmful chemicals; biologically and chemically stable; and slow to decompose. Cost should also be considered when comparing one material with another.

GROWING MEDIA

Soil

Natural soil alone is seldom the best choice for indoor use, but with the addition of certain materials a suitable medium may be achieved.

Depending on the nature of the parent material, soil may not be adequately drained or aerated. When using natural soil, its texture and structure are prime concerns.

Soil Texture.   Texture refers to the kinds of particles present. For potting soils, the sand, silt, and clay factions are important. Sand particles range in size from 2.0 to 0.05 mm in diameter. They have a small surface area and are not chemically active. Thus, they retain little water and few nutrients and produce soils that are well drained but not very fertile.

Clay particles are less than 0.002 mm in diameter, the smallest soil component. They are, however, the most important. Clay particles are negatively charged and react with other charged particles (cations or positively charged particles) in the soil. Cations that may include many of the essential plant nutrients are held on the surface of the clay and may be exchanged for other ions in the soil solution or with the roots. Thus, essential elements can be held in the soil by the clay and made available for eventual use by the plant. Most numerous on the exchange sites are the ions of calcium, magnesium, hydrogen, sodium, potassium, aluminum, and ammonium. The proportions of ions in the soil are constantly changing due to dissolving minerals; plant absorption; the addition of lime, gypsum, and fertilizers; and leaching.

Clay particles have a large surface area and because of their small size, produce a soil with many small pores. Such soils drain slowly, so that waterlogging and poor aeration may be a problem in soils high in clay. Many clay soils expand and contract with wetting and drying. Drying causes shrinkage of the soil from the walls of the planter, making subsequent watering difficult. Such soils may also become hard and crack if dry.

Silt particles are intermediate between sand and clay, ranging in diameter from 0.05 to 0.002 mm. Silty soils have inadequate drainage.

Loams are soils which are combinations of sand, silt, and clay, and do not exhibit the dominant physical properties of any of the three groups. Sandy loams or loams (Figure 7.1) are best for use in containers, as they have a balance of pore sizes to facilitate drainage and aeration.

Soil Structure.   Soil structure, the arrangement of the particles, is another important physical aspect. Sand tends to function as individual particles, whereas clay and silt form aggregates. Aggregates are secondary particles composed of many soil particles bound together by organic substances, iron oxides, carbonates, clays, and/or silicates. A soil that is well aggregated tends to have numerous large pores and is well

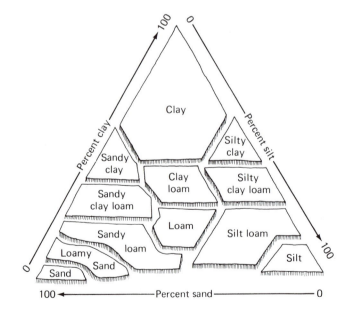

Figure 7.1   Texture triangle showing the percentages of sand, silt, and clay in each soil textural class. (Courtesy of USDA.)

drained. Many factors influence soil aggregation. Products of organic matter decay will cement particles together, increasing aggregation; and working soil that is too wet destroys them. Calcium ions help aggregate, while sodium ions (Na^+) destroy soil structure by breaking up aggregates. Softened water may be high in Na^+, thus adversely affecting soil structure.

Natural soil as a component of any growing medium for container plants has excellent nutrient retention and is a source of some trace elements. It has a greater bulk density (is heavier) than many other materials and may support the plant better. If aggregation is enhanced to provide good drainage and aeration, a satisfactory growing medium can be achieved using natural soil. To improve soil structure, organic material and coarse aggregates are added.

Addition of Organic Material. Organic material is often added to the soil to loosen it, provide channels for the drainage of water and entry of air, and to improve water retention. The presence of organic matter prevents close packing of the soil particles, promoting aggregation. It usually improves the cation-exchange capacity. In sands, moisture retention is improved, and in clays, pores are provided and compaction reduced. In selecting any organic soil additive, it is desirable to use materials that are coarse-textured and resistant to decay. Although decomposing organic materials do release some essential minerals, such as nitrogen, phosphorus, and sulfur, they should not be considered a primary source of nutrient. In fact, depending upon the speed of decomposition, they may actually cause nitrogen deficiency. Decomposition is caused by soil microorganisms, plants that also require nitrogen. With rapidly decomposing organic matter, there is competition for the available nitrogen, and a deficiency may occur.

There are many organic materials that may be incorporated into natural soil. When selecting soil amendments, one must consider not only their effectiveness in improving soil structure, but also their uniformity, sterility, chemical composition, ease of use, reliability of supply, and cost, and select those materials that do the job most economically.

Sphagnum Peat Moss. Peat moss is readily available and perhaps the most frequently used organic additive. It is a brown, fibrous material, composed of partially decomposed sphagnum moss. Peat is dug from bogs in Canada, Poland, West Germany, and other areas.

Sphagnum peat is easy to handle and store, and light in weight when dry. Depending on the source, it will absorb up to 20 times its weight in water which is held under low tension and available to the plant, thus improving the water retention of the soil. Peat can be dried and rewetted again and again without losing its water holding ability. Dry peat may be difficult to wet, and the use of a liquid or granular wetting agent such as Aqua-GRO™ is advisable. Peat moss is resistant to decomposition and generally free of weed seeds, pests, and pathogens. It is acid in reaction, with a pH usually less than 5.0, and may require the addition of ground limestone to raise the pH to a satisfactory level for many foliage plants. Peats vary in soluble-salt content.

As a component of the growing medium, sphagnum peat increases the cation-exchange capacity, thereby holding elements in the medium for absorption by the plant. The moss itself is low in minerals.

There are two types of sphagnum peat. Light-colored or young peat moss is more fibrous and elastic and fosters better aeration and drainage than the dark brown moss, which is more decomposed. Brown peat has more organic matter, holds more water and nutrients, and may require more sand, perlite, bark, or other aggregate to aid drainage than the younger moss. Either is a satisfactory component of the growing medium when properly managed.

Sphagnum peat is available in particles of various sizes. For use as a growing medium, the horticultural grade is recommended. When purchasing peat, various-sized packages are available, ranging from 0.2-ft^3 bags of loose peat to 4-ft^3 compressed bales. Purchase those sizes that will yield the quantity needed at the lowest price. A compressed bale of peat will "fluff up" and yield at least 50% more volume as a soil component than indicated on the bale. For example, a 4-ft^3 bale yields 6 to 7 ft^3 of peat, light-colored yielding more than dark.

There are other kinds of peat available, including reed-sedge and humus peats. These are the black materials usually sold in bags or by the truckload. Such peats are too decomposed and not satisfactory for use as a component of the growing medium. In fact, they may restrict drainage and aeration, rather than improve it.

Wood Residues.   Various wood residues may be used satisfactorily as components of growing media. Some may decompose rapidly, causing nitrogen deficiency during the first few weeks.

Sawdust is the most common and widely used wood residue. If readily available and inexpensive, it may be incorporated into the medium to improve the physical condition. Sawdust is fairly durable, with that from hardwoods (oak, maple, etc.), which are high in cellulose, decaying more rapidly than sawdust from softwoods (pine, spruce, etc.), which contain little cellulose. One should avoid using sawdust from incense cedar and walnut, as it is toxic to the plants. Redwood sawdust may also be toxic.

Sawdust has a high carbon/nitrogen ratio, which is conducive to the growth of microorganisms in the warm, moist growing media used indoors. Since these microorganisms are plants which also require nitrogen for growth, there is competition for the available nitrogen, and the foliage plants may become nitrogen-deficient. Applications of supplemental nitrogen during the first few months will eliminate this problem, but add an additional cost which must be considered in evaluating sawdust as a medium component. The saprophytic decay-causing fungi of sawdust are not harmful, but their growth may impede water penetration, and they are not attractive in planters. Pasteurization (see later) of sawdust to eliminate pathogenic organisms is essential.

Wood chips are another residue, but the particles are too large for use as a medium. They may be used quite satisfactorily as a mulch, however.

Bark may be used as a mulch or a component of the medium. Fir, cedar, redwood, pine, and hemlock barks are satisfactory amendments, provided that they are incorporated in proper proportions with other mix ingredients and that the particle size is correct. A particle size of ⅛ to ¼ in. or less is most desirable. Larger particles are used for mulching. Milled pine bark is usually hot-air-dried and may be difficult to wet. Composting overcomes the problem. Wettability improves as the moisture content increases. Because a bark substrate may cause water stress at the time of planting, the medium must be sufficiently moist prior to use. As with sawdust, softwoods are more durable than hardwoods. Nitrogen depletion is not as severe with barks when compared to sawdust of the same species. Bark has an acid pH and low cation-exchange capacity.

Compost.   A variety of composted organic materials are available for use as a substrate component. These include municipal refuse (garbage), wastewater biosolids composted with wood chips, and fermentation residues. Used as a medium component, they may improve drainage and aeration, water-holding, cation-exchange capacity, and add nutrients, especially trace elements.

Before using composts, the chemical composition should be determined as they may be high in soluble salts, including heavy metals. Evaluate their usefulness on a small scale before using large batches. Composts are used to replace some of the peat or bark in the medium. Depending upon chemical composition, 10–40% compost, by volume, is acceptable.

Addition of Coarse Aggregates.   Coarse aggregates are added to the growing medium to improve drainage and aeration. Satisfactory coarse aggregates include sand, vermiculite, perlite, calcined clay, and polystyrene beads.

Sand.   Sand is the least expensive and heaviest of the materials available. Sharp sand with particles 0.5 to 2.0 mm in size should be used in sufficient volume; otherwise, drainage and aeration may not be improved. Do not use beach sand, as it is high in soluble salts and may injure plants.

Sand provides no minerals for plants, nor does it have any cation-exchange capacity. It may be highly contaminated with pathogens and must be pasteurized.

Vermiculite.   Vermiculite is a mica compound that has been heated to 1400°F in processing and is therefore sterile. It has a high cation-exchange capacity with minerals being held on the surface and between the particle layers. In addition, vermiculite contains about 6% potassium and 20% magnesium. The pH is usually about 7; however, African ores may be very alkaline with a pH of 9 to 10. The water-holding capacity of the medium is increased with the incorporation of vermiculite, and the bulk density decreased.

Vermiculite is not as durable as sand or perlite. It is easily compressed, and the particles break apart readily if they are handled when wet.

The material is usually purchased in 4-ft³ bags. Various grades are available, with No. 2, 3, or 4 being satisfactory for incorporation into growing media. Insulating grades should be avoided, as the particles have been made waterproof.

Perlite. Perlite is produced from a siliceous volcanic rock which "pops" when heated to 1600°F, resulting in white particles 4 to 20 times larger than the original. Perlite particles have an irregular surface with many cavities and an interior of closed air bubbles. Water is held on the surface. Perlite has no cation-exchange capacity.

As a component of the growing medium, perlite does not deteriorate, although it may fragment. It is light in weight (7 lb/ft^3), safe to handle, odorless, and sterile. Perlite is extremely dusty when dry, but when moist it is clean and easily handled.

As a coarse aggregate, perlite improves drainage and aeration. It retains three to four times its weight in water but even when moist is less dense than water, and floats. The white color assures uniform mixing. Perlite has a pH of 6.5 to 7.5.

Perlite is sold in 4-ft^3 bags and is easily stored. A medium-to-coarse horticultural grade should be used.

Calcined Clay. This aggregate is formed from montmorillinic clay fired at 1300°F. The resulting particles are porous and retain water. Calcined clays have a moderate cation-exchange capacity. The particles are very stable and extremely durable. Calcined clay is heavy, weighing five to six times more than perlite or vermiculite, but with only 60% of the weight of sand.

Polystyrene Beads. A recently available aggregate, polystyrene beads are odorless, neutral, and resistant to decay, but may disintegrate in pasteurization. Being plastic, they hold no water, and will float. They do not contribute to the cation-exchange capacity. In addition, they are charged with static electricity, especially when dry, and cling to each other and any surfaces they contact. Mixed with other components of the medium, they do improve drainage and aeration, however.

Rockwool. Rockwool is an inert, sterile growing medium made from the mineral basalt heated to high temperatures and extruded into fibers. The shredded form is suitable for use alone or as a component of soilless mixes. Water-absorbent forms increase water-holding while fostering drainage and aeration; do not use insulation grades. The material is completely inert, and provides no nutrients, has no notable cation-exchange capacity, and does not decompose.

Hydrophilic Polymers. Another amendment that may be incorporated when preparing the medium is a water-absorbing compound called a hydrophilic polymer. Capable of absorbing 40 to 800 times their weight in water, these granules or powders form gel-like particles when hydrated. They increase water-holding capacity while simultaneously improving drainage and aeration. Fertilizer is also absorbed. Effectiveness is related to the type of gel and growing medium used.

The roots grow into and around the particles which release the water and fertilizer to the plants over time. Irrigation frequency is reduced, and nutrition may be more uniform. Nutrient leaching is reduced; pH is not affected. Gels can be rehydrated and remain active for several years.

Use of hydrophilic polymers as additives to highly organic container media is controversial. Research has shown that concentrations of fertilizer salts typically used in commercial production of foliage plants may restrict the potential benefits of polyacrylamide gels. Also, high-quality peat-lite and other soilless mixes already have excellent water-holding capacity and may not benefit from the addition of a gel. Hydrophilic polymers may reduce the frequency of irrigation. Evaluate the effectiveness and the cost of the material compared with the benefits before proceeding with large-scale use. Available hydrophilic polymers include SuperSorb™ and Soil Moist Plus.™

Soil Mixtures

Having an understanding of the nature of soil and the difficulties one may encounter using natural soil alone, one can create a suitable growing medium by mixing components. Most foliage plants, except cactus, prefer a highly organic, acid growing medium. The following mixes should be satisfactory.

Heavy soil (clay or clay loam)

1 part soil : 3 parts sphagnum peat : 2 parts coarse aggregate (by volume)

Medium-textured soil (silt loam or sandy clay loam)

2 parts soil : 3 parts sphagnum peat : 2 parts coarse aggregate (by volume)

Light soil (sandy loam)

2 parts soil : 3 parts sphagnum peat (by volume)

To each bushel of mix, incorporate gypsum or limestone for calcium as specified by a soil test.

In preparing the growing medium, thorough mixing is essential. Small quantities may be mixed by hand using a shovel; shredders or cement mixers may be used for large volumes.

Where small volumes of growing media are used, as in many plant shops, one may purchase pasteurized soil in 25- or 50-lb bags. This is field soil or muck and will have to be amended by the addition of peat and vermiculite or perlite. Other types of prepared potting soils may not have the desired physical characteristics and will prove unsatisfactory if used directly from the bag. Purchased soils may vary from bag to bag and be high in soluble salts, have the wrong pH, or have a nutrient imbalance, and should be tested before they are used (Table 7.1).

PASTEURIZATION OF THE MEDIUM

Whenever natural soil, sand, and wood residues are used as components of the growing medium, insects, weed seeds, and pathogens may be present. They must be destroyed before the medium is used in the indoor garden. Pasteurization is the final stage in substrate preparation. Start with a loose, damp medium to which all the various amendments have been added, and proceed to destroy potentially harmful components with either heat or chemicals.

TABLE 7.1 Analysis of Some Commercially Prepared Growing Media[a]

Growing Medium	pH	Soluble Salts	Nitrate-N (ppm)	P	K	Ca	Mg	C.E.C. (meq/100 g)
Black Magic African Violet Soil	4.5	44(L)	404	M	L	M	VH	32.5
Black Magic All Purpose Planter Mix	5.5	65(L)	405	XS	XS	M	VH	35.9
Black Magic House Plant Mix	6.5	32(L)	102	L	H	M	VH	25.5
Black Magic House Plant Mix	6.8	22(L)	18	VL	L	L	XS	28.5
Burpee All Purpose Potting Mix	6.6	600(XS)	404	VH	H	H	VH	54.6
Fertilife Potting Soil	6.5	650(XS)	404	H	H	H	H	59.7
Gardener's Potting Soil	6.6	560(XS)	404	H	VH	VH	VH	55.3
Hoffman Potting Soil	6.7	42(L)	10	VL	L	VH	L	26.4
K-Mart Potting Soil	6.7	130(M)	237	L	L	VH	H	52.6
Nature's Balance Potting Soil	6.2	725(XS)	809	XS	XS	L	XS	63.0
New Era African Violet Soil	4.9	130(M)	214	MH	M	ML	VH	25.8
New Era Potting Soil	4.5	140(MH)	178	L	L	L	VH	26.5
New Era Potting Soil	6.1	100(M)	8	L	L	M	XS	35.7
Swiss Farms African Violet Soil	6.1	360(XS)	404	L	L	VH	ML	47.3
Swiss Farms Planting Mix	6.1	220(XS)	287	XS	L	VH	M	49.5
Swiss Farms Potting Soil	6.4	180(H-XS)	214	MH	L	VH	L	43.1
Swiss Farms Potting Soil	6.1	150(H)	108	L	L	H	M	30.3
Terra-Lite Redi-Earth	5.8	60(L)	404	H	MH	M	VH	38.1
Worm Rich	5.6	180(XS)	342	H	XS	M	VH	27.0

[a]VL, very low; L, low; ML, medium low; M, medium; MH, medium high; H, high; VH, very high; XS, excess.

Heat

Heat is an effective means of destroying pathogenic organisms and weed seeds. Most pathogens are killed at 140°F for 30 min and weed seeds at 180°F. Dry heat or aerated steam can provide 180°F, while raw steam heats the medium to 212°F.

Dry Heat Pasteurizers. Dry heat produced by electrical resistance units or a kitchen oven may be used to pasteurize the growing medium. Soil is placed within the unit (Figure 7.2), and heat is conducted and convected through the substrate. The units are thermostatically controlled and go off after a set time. In a dry substrate the temperatures must be higher than desired so as to reach the outer extremities. The medium in contact with the heat source will get very hot, and the organic matter may be destroyed. In moist media, steam may be generated, causing the overheating problems associated with pasteurization using raw steam. Electric pasteurizers are available for quantities of soil ranging from 1 ft³ to ⅛, ¼, and ½ yd³.

Steam. The most frequently used form of moist heat for pasteurization is steam. Pasteurization with steam, as with electrically generated heat, can be done at any time, is nonselective, is not toxic to plants in the vicinity, and is not toxic to people, although the heat may be a problem. Once pasteurized, the medium may be used as soon as it cools.

Bulk sterilizers are ideal for steam pasteurization of prepared substrate. A bin constructed of wood, metal, or cinder blocks open at the top and one side is satisfac-

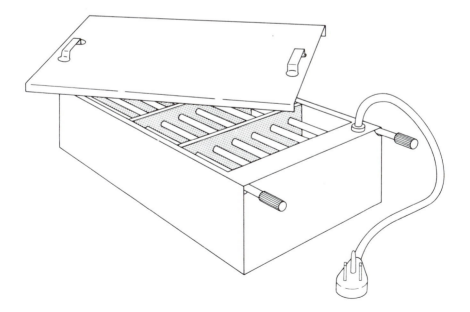

Figure 7.2 Electric soil pasteurizer.

tory (Figure 7.3). A grid of pipes perforated with ³⁄₁₆-in. holes spaced 12 in. apart on the underside delivers steam from the boiler. Several tiers may be necessary, as the medium should not be more than 6 in. from a steam outlet. The medium is placed in the bin and covered with a neoprene or rubber tarpaulin anchored along the edges. The steam is turned on and allowed to flow until the coldest portion maintains > 180°F for 30 min. Actually, the steaming temperature will be 212°F. If the entire steaming process takes more than 1½ hr, the volume of soil is too large in relation to the volume of steam, and less medium should be treated at one time. Excessively long steaming may destroy soil structure and release phytotoxic chemicals. Other types of bulk sterilizers include special boxes, dump-truck bodies, and portable benches.

Raw steam heats the soil to 212°F, which is far in excess of the 160°F necessary to destroy most weed seeds and pathogens. Except for a few resistant organisms, the medium is sterilized. Various problems may result.

High-temperature steaming increases the solubility of many compounds, especially phosphates, potash, manganese, zinc, iron, copper, and boron. Thus, the soluble salts are increased and may cause injury. To minimize this aspect, start with a low-fertility soil and do not incorporate fertilizer except superphosphate and lime before pasteurization.

Manganese is released upon steaming and may become toxic. Raising the pH and reducing the steam time should minimize the problem.

Ammonia buildup is a frequent problem following steaming. The nitrifying bacteria are destroyed in steaming, but the more resistant ammonifying bacteria are not. After steaming, the ammonifying bacteria become active, producing ammonia from organic compounds containing nitrogen. Since the nitrifying bacteria that convert the ammonia to nitrites and nitrates are not present, toxic levels of ammo-

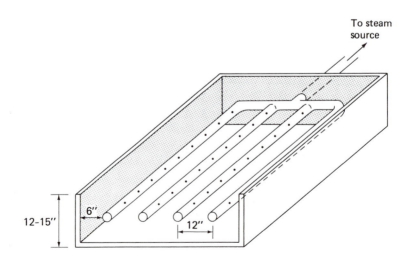

Figure 7.3   Bulk pasteurizer using steam.

nia may build up in the first 2 to 3 weeks following steaming. Gradually, the populations of nitrifiers are reintroduced and a balance is restored. Avoiding the use of liming materials plus frequent irrigations will reduce the problem.

Since most of the microflora are killed with raw steam, there is no competition and, if a pathogen contaminates the sterile soil, the pathogen will spread very rapidly. The danger of contamination may be avoided by sanitation and by steaming at a lower temperature using aerated steam. The use of soil fungicides after steaming will also control new infestations of pathogens and is recommended.

Aerated steam is created when air is injected into the steam line. With a given weight ratio of air to steam, the temperature of the steam–air mixture is reduced to a desired level. Aerated steam at 140 to 160°F will destroy the pathogens in the soil. It reduces the release of manganese and poststeaming toxicity. A portion of the saprophytic microflora survives and provides competition and/or antagonism should pathogens be introduced. The quantity of steam required is reduced, thereby saving energy and reducing the cost of the operation. Because the temperatures are lower, media pasteurized with aerated steam cool faster and can be used sooner than when raw steam is used. Equipment is available for injecting air into the steam line.

Hot Water.   Hot-water treatment is another method of pasteurization. Because large quantities of water are required, an extremely wet, difficult-to-dry medium results. The temperature cannot be kept uniform. This method has nothing to recommend its continued use.

Solar Heating.   In warm regions, solar heating will kill weed seeds and many plant pathogens in potting media. Fill clear plastic bags with prepared, moist medium, seal, and place in the sun. Turn the bags weekly. In three or four weeks, the medium can be used. During the hottest time of the year, temperatures in the bag can exceed 120°F (51°C).

Chemical Fumigants

Chemical fumigants have been developed which may be used to destroy pests, diseases, and seeds. Unless large quantities of growing medium are involved and steam is not available, chemicals are not usually used.

The materials available include formaldehyde and chloropicrin. They are highly toxic to plants and people, and extreme caution must be taken in using them. They are not lethal to a broad spectrum of organisms. Fungal pathogens are more resistant than insects, nematodes, or weed seeds. Perhaps the greatest disadvantage to using chemicals is that the medium cannot be used until they have been dissipated by evaporation. This may take several weeks.

SOILLESS MEDIA

When natural soil is used as part of a growing mix, standardization of the medium is difficult, as soils vary depending upon their source. Natural soil is becoming very expensive. For these reasons, soilless media have become the norm in interior plantscapes. They may be purchased or made at home.

Types

U.C. Mixes. Developed at the University of California, U.C. Mixes are composed of fine sand (0.5 to 0.05 mm) and sphagnum peat. Five basic mixes are created by varying the amount of the components. Two commonly used formulations are 1 part sand : 1 part peat by volume and 1 part sand : 3 parts peat by volume. U.C. Mixes are standardized, relatively inert, easily prepared, and well drained and aerated. In selecting the sand, dune sands with round particles should be used, as block-shaped particles may cause packing. The medium is heavy and low in nutrients at the outset but has good nutrient retention. A fertilizer program must begin immediately, or a base fertilizer may be incorporated into the mix. Pasteurization is essential.

Peat-Lite Mixes. Composed of sphagnum peat moss plus vermiculite and/or perlite, the peat-lite mixes were developed at Cornell University. The ingredients are readily available in standard and uniform quality, so that the mix may be easily prepared and duplicated. Peat-lite mixes may be purchased premixed in bags or bales. Remixing prior to use to homogenize the medium is recommended. The medium is light in weight and sterile and has the proper chemical and physical properties for container-grown plants. Some nitrogen is available, but it must be supplemented. Costs are generally competitive with other growing media.

There are certain disadvantages to peat-lite mixes, but these are easily overcome. Because of the high water-holding potential, greater skill is required in judging when to water peat-lite mixes than is required for media containing natural soil. They are not watered as frequently. Initially, the peat may be difficult to wet, but this problem may be eliminated by wetting before mixing or using, or by applying a nonionic wetting agent such as Aqua-GRO immediately after planting. More than one application may be necessary to thoroughly wet the mix. Warm water is suggested for the initial watering to further aid wetting.

High nitrogen in the base fertilizer, incorporated during mixing, may raise the level of soluble salts. In time, trace-element deficiencies will occur, requiring additions of microelements to the fertilizer. Because of the low bulk density of peat-lite mixes, plants may topple easily.

In preparing peat-lite mixes, care should be exercised so as not to contaminate the components. Mix on a clean sheet of plastic. Two peat-lite formulations are commonly used for interior plants. The foliage mix described in Table 7.2 is for fine-rooted plants requiring a medium with high moisture retention, including *Aphelandra*, *Begonia*, *Justica* (*Beloperone*), *Cissus*, *Coleus*, ferns, *Ficus*, *Hedera*, *Solierolia* (*Helxine*), *Maranta*, palms, *Pilea*, *Sansevieria*, and *Tolmiea*.

The epiphytic mix described in Table 7.3 incorporates composted bark with the peat and perlite, and is for plants with coarse, tuberous, or rhizomatous roots requiring good drainage and aeration with some drying between irrigations. This

TABLE 7.2   Peat-Lite Foliage Mix

Component	Quantity per Cubic Yard	Quantity per Bushel
Sphagnum peat (½-in. mesh)	½ yd³ (11 bu)	½ bu
Horticultural vermiculite (No. 2)	¼ yd³ (5½ bu)	¼ bu
Perlite (medium)	¼ yd³ (5½ bu)	¼ bu
Ground dolomitic limestone	8¼ lb	8 level Tbls
20% superphosphate, powdered	2 lb	2 level Tbls
10-10-10 fertilizer	2¾ lb	3 level Tbls
Iron sulfate	¾ lb	1 level Tbls
Potassium nitrate (14-0-44)	1 lb	1 level Tbls
Fritted trace elements	2 oz	Omit
Granular wetting agent	1½ lb	3 level Tbls

TABLE 7.3   Peat-Lite Epiphytic Mix

Component	Quantity per Cubic Yard	Quantity per Bushel
Sphagnum peat (½-in. mesh)	⅓ yd³ (7⅓ bu)	⅓ bu
Douglas (red or white) fir bark (⅛–¼ in. size)	⅓ yd³ (⅓ bu)	⅓ bu
Perlite (medium)	⅓ yd³ (7⅓ bu)	⅓ bu
Ground dolomitic limestone	7 lb	8 level Tbls
20% superphosphate, powdered	4½ lb	6 level Tbls
10-10-10 fertilizer	2½ lb	3 level Tbls
Iron sulfate	½ lb	1 level Tbls
Potassium nitrate (14-0-44)	1 lb	1 level Tbls
Fritted trace elements	2 oz	Omit
Granular wetting agent	1½ lb	3 level Tbls

medium is recommended for African violets, *Aglaonema*, *Aloe*, bromeliads, cactus, *Crassula*, *Dieffenbachia*, *Epipremnum*, *Episcia*, *Hoya*, *Monstera*, *Philodendron*, *Syngonium*, and *Peperomia*.

Commercially prepared peat-lite mixes are available and include Jiffy-Mix™, Pro-Mix™, Redi-Earth™, Metro Mix 220™, and various Southland Mixes.

Bark Mixes.   A soilless medium containing bark, such as a mixture of 3 parts composted pine bark : 2 parts sand : 1 part sphagnum peat by volume is excellent. Because pine bark is acid, 10 lb of dolomitic limestone and 3 lb of hydrated lime per cubic yard should be incorporated in preparing the mix. Addition of minor elements either in mixing or in regular applications is an absolute necessity. Fertilizer programs for bark media are similar to those for other substrates, except that the nitrogen should come from a nitrate source such as calcium nitrate or potassium nitrate and not from ammonium nitrate. A 15-15-15 fertilizer is excellent for bark-containing mixes; 20-20-20, which contains a higher ratio of ammonium ions, should be avoided. Commercially prepared bark mixes such as various Metro Mixes, Pro-Mix NX™, and Fafard Nursery Mix™ are readily available and will give excellent results.

There are a large number of soilless products available commercially. They vary in content from single components such as peat moss, perlite, or vermiculite to mixtures of two to five or six ingredients, and may have specific uses. Some have additives such as fertilizer or a wetting agent. Various-sized packages are available. Interior plantscapers must make their own decisions as to the best growing medium to use.

Storage

Storage of soilless mixes is relatively easy, particularly when they are purchased in bags or bales, but certain precautions should be practiced.

1. Keep the mix away from herbicides. Some herbicides are vapor-active and may be absorbed by the medium and cause injury later.
2. Store under cover. Sunlight can deteriorate the plastic bags and allow water to penetrate.
3. Store off the ground, so that disease organisms cannot enter through tears that may occur in the bags.
4. The longer in storage, the drier the medium becomes. Add 5 to 7 gal of water to the bag the night before using in order to wet the mix.

Costs

Before selecting any medium for the interior plantscape, several factors should be considered. All costs should be accounted for, including costs of the components, labor for mixing and handling, equipment, pasteurizing, fertilizer and other additives, and storage. The availability of the components and their uniformity from one time to another, and one's ability to properly manage the substrate must also be taken into account. Considering all the factors involved, the cost of mixes containing soil may exceed soilless media purchased ready to use.

MOISTURE AND AERATION

The amount of moisture and air present in the medium in a drained container is a function of the pore space. A proper balance of coarse, medium, fine, and very fine pores must be present to provide adequate moisture and air.

Not all of the water in the medium is available to the plant. Gravitational water drains quickly from the large pores as a result of the force of gravity and is only slightly available for absorption by the roots. Following free drainage, a certain amount of water will be held in the medium. In a field with deep soil, this is called field capacity; but container media, being shallow, hold water in excess of field capacity, so that container capacity is perhaps a more accurate term to use. Water between container capacity and the wilting point is available for plant use. The wilting point is the amount of water in the medium when plants will not recover from wilting unless additional water is added. Water held with tensions below the wilting point is not available to the plant.

As the depth of the medium is reduced, the tension force of gravity acting on the water column in the medium is also reduced. Water cannot cross the soil/air interface at the bottom of the container until the weight of the water exceeds atmospheric pressure at the interface and the adhesive forces between the medium particles and the water. Thus, the pores at the base of the container remain filled with water. In shallow containers, the saturated zones occur closer to the surface (Figure 7.4) and may result in root injury. To alleviate the situation, the depth of the soil should be increased by using deeper containers. Medium structure should be improved by increasing the number of large pores in the medium.

The influence of medium depth on drainage is easily demonstrated using an ordinary household sponge saturated with water. Lift the sponge and hold it with the smallest dimension perpendicular to the earth. A small amount of water drips from the lower edge. Turn the sponge so that the next longer side is vertical; additional water will be lost. Finally, place the longest side perpendicular to the earth. Because the column of water has been increased in height, still more water is lost from the sponge.

Pores in the medium that are not filled with water are filled with air. Oxygen is essential for respiration in the cells of the roots and is absorbed directly from the atmosphere of the growing medium. Without oxygen, the cells of the roots suffocate, causing some or all of the root system to die. Without a healthy root system, the entire plant will be adversely affected, producing typical symptoms of overwatering. Death of the plant may ensue.

In an adequately drained growing medium, the roots will be respiring, using O_2 and producing CO_2. Because air between the medium and the atmosphere is

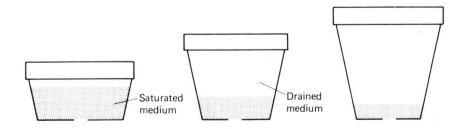

Figure 7.4 Influence of container depth on the drainage of a growing medium. Deeper substrates drain better than shallow media of the same kind.

exchanged by diffusion at a relatively slow rate, the levels of O_2 and CO_2 in the medium will differ from that in the ambient air. O_2 will be lower and CO_2 higher. Inhibition of root growth occurs in many plants if the O_2 in the soil air is less than 10% and CO_2 greater than 5%. These extremes are difficult to obtain in containers, so that the composition of the atmosphere of the growing medium should not be a problem.

SOIL TESTING

Testing growing media before planting and periodically thereafter assesses the mineral status of the medium, thereby assisting interior landscapers in the evaluation of their fertilizer program and aiding in the diagnosis of nutrition-related plant disorders. Testing is performed for a nominal fee by the Agricultural Extension Service and private laboratories. County agricultural agents can provide information on the specific soil-testing procedure in your state.

To be meaningful, the sample must be representative of the medium in question. To assess general nutrition, a composite sample from several containers is desirable. In the case of a troubled plant, the medium sample should be taken from the troubled area only. The questionnaire accompanying the sample should be filled in as completely and accurately as possible.

The sample one collects must be representative of the medium, and need not destroy the plant. In small pots, remove the plant from the pot. Using a sharp knife with a serrated blade, cut a pie-shaped wedge from top to bottom through the root ball. Put fresh medium in the cut area and return the plant to its pot. Take the wedge and cut off and dispose of the top $1/2$ to 1 in. Use the remainder for the sample. Thus any top-dressed fertilizer and soluble-salt residues are removed. Repeat with additional pots if a composite sample is to be collected.

For large pots, a soil sampling probe can be used. Remove the top inch of medium before inserting the probe, and be sure to collect a sample to the bottom of the container. Repeat in other planters for composite samples. Air dry the samples before testing. If tests are to be run in-house, carefully follow the directions that came with the equipment. If they are to be sent to a laboratory, send the amount needed to run the tests as specified by the lab.

Following analysis of the growing medium, a report showing results and recommendations will be issued. The information provided usually includes the soil pH, the soluble-salt level (see Chapter 8), and the levels of nitrogen, phosphorus, potassium, calcium, and magnesium. The cation-exchange capacity may also be given. Recommendations for correcting any weaknesses, or excesses, are part of a soil test report. A soil test report for Pennsylvania is shown in Figure 7.5.

pH

The pH of the growing medium should be of concern for the indoor gardener. pH is an index of the acidity or alkalinity of the medium and ranges from 1.0, which is

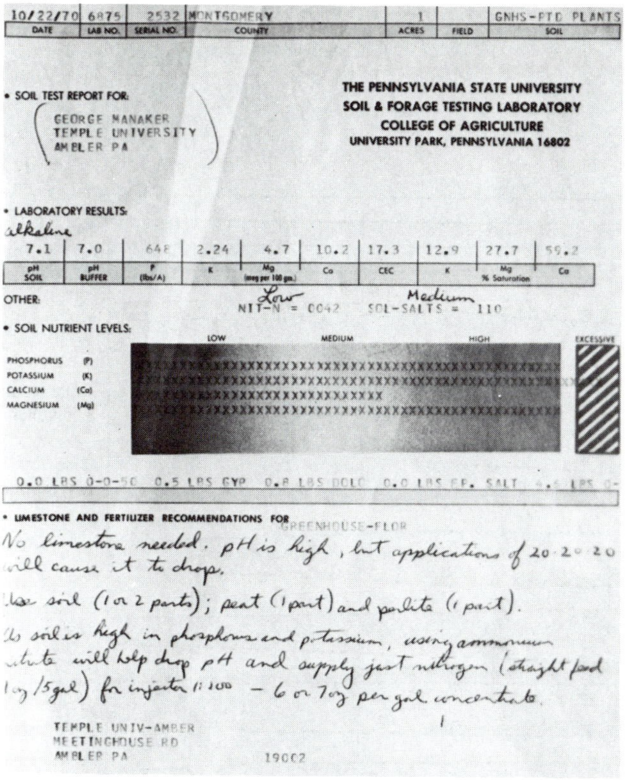

10/22/70	6875	2532	MONTGOMERY	1		GNHS-PTD PLANTS
DATE	LAB NO.	SERIAL NO.	COUNTY	ACRES	FIELD	SOIL

• SOIL TEST REPORT FOR:

GEORGE MANAKER
TEMPLE UNIVERSITY
AMBLER PA

THE PENNSYLVANIA STATE UNIVERSITY
SOIL & FORAGE TESTING LABORATORY
COLLEGE OF AGRICULTURE
UNIVERSITY PARK, PENNSYLVANIA 16802

• LABORATORY RESULTS:

alkaline

pH SOIL	pH BUFFER	P (lbs/A)	K (meq per 100 gm.)	Mg (meq per 100 gm.)	Ca (meq per 100 gm.)	CEC	K (% Saturation)	Mg (% Saturation)	Ca (% Saturation)
7.1	7.0	648	2.24	4.7	10.2	17.3	12.9	27.7	59.2

OTHER: NIT-N = 0042 (Low) SOL-SALTS = 110 (Medium)

• SOIL NUTRIENT LEVELS:

0.0 LBS 0-0-50 0.5 LBS GYP 0.8 LBS DOLO 0.0 LBS EP. SALT 4.5 LBS 0-

• LIMESTONE AND FERTILIZER RECOMMENDATIONS FOR GREENHOUSE-FLOR

No limestone needed. pH is high, but applications of 20-2-20 will cause it to drop.

Use soil (1 or 2 parts); peat (1 part) and perlite (1 part).

As soil is high in phosphorus and potassium, using ammonium nitrate will help drop pH and supply just nitrogen (straight feed 1 oz/5 gal) for injector 1:100 – 6 or 7 oz per gal concentrate.

TEMPLE UNIV-AMBLER
MEETINGHOUSE RD
AMBLER PA 19002

Figure 7.5 Soil test report from the Soil and Forage Testing Laboratory at The Pennsylvania State University.

extremely acid, to 14.0, which is alkaline. A pH of 7.0 is neutral. Foliage plants generally prefer a slightly acid medium. For mineral soils, maintain a pH of 6.5 to 6.8; organic soilless media should have a pH of 5.2 to 5.5. The pH of the medium is important because it will affect the availability of all the mineral elements needed by the plant, as shown in Figure 7.6. In the pH range recommended for most foliage plants, all essential minerals should be available.

It is important to know the pH of the medium before it is used. The pH of the medium also changes during use and is influenced by its components, the fertilizers used, and the accumulation of residues from irrigation water. Periodic testing is desirable. Complete soil tests will include pH. The pH may be rapidly determined on site using an inexpensive digital pH meter or pH testing papers, which are available from greenhouse supply firms. Careful adherence to directions will produce an accurate test. Low-cost probe pH meters are sold by many suppliers. They have been shown to overestimate pH and appear to be too inaccurate for horticultural use.

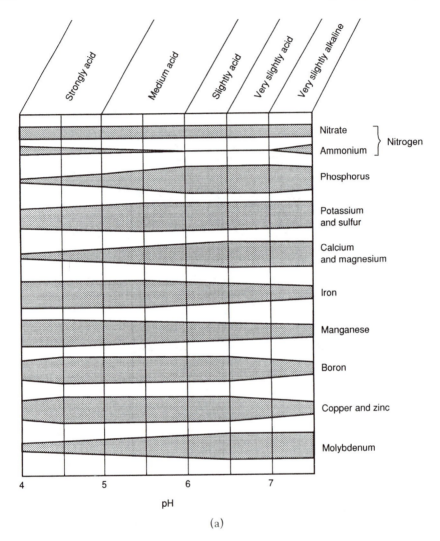

(a)

Figure 7.6 (a) Influence of pH on the availability of minerals in a soil-based growing medium. (From J. S. Koths et al., *Nutrition of Greenhouse Crops,* University of Connecticut Bulletin 76-14, 1976, p. 2.) (b) Influence of pH on the availability of minerals in a soilless medium, W. R. Grace Metro Mix 300. (Adapted from J. C. Peterson, "Effects of pH upon Nutrient Availability in a Commercial Soilless Root Medium Utilized for Floral Crop Production," *Ornamental Plants–1982, A Summary of Research,* Research Circular 253, Ohio Agricultural Research and Development Center, Wooster, Ohio.)

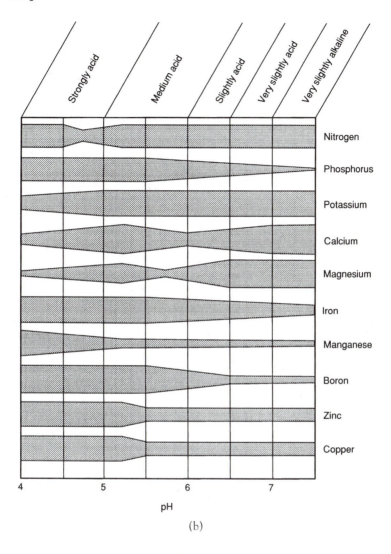

(b)

Figure 7.6 (continued)

Changing the pH of a growing medium can take time because many of the materials used are slow-acting; therefore, growing media should be mixed well in advance of use. Rate of change varies with medium type. Highly organic media and media high in clay require more corrective material and time than sandy media of low organic content. Changes are more rapid in warm, moist, well-aerated substrates.

Should a correction of pH be required, the following recommendations for initial treatments are made. The rates of application are minimal, and subsequent monitoring will determine if more is needed.

To change the pH to 6.0 to 7.0 in a 6-in. pot if the pH is

3.5–4.0	use 4 tsp ground limestone
4.0–4.5	3 tsp
4.5–5.0	2 tsp
5.0–5.5	1½ tsp
5.5–6.0	1 tsp

For more acid plants, to change the pH to 5.0 to 5.5 in a 6-in. pot if the pH is

3.5–4.0	use 2 tsp ground limestone
4.0–4.5	1 tsp
4.5–5.0	½ tsp
5.5–6.0	¼ tsp finely ground sulfur
6.0–6.5	½ tsp
6.5–7.0	¾ tsp
7.0–7.5	1 tsp

A lime-water drench will raise the pH more rapidly than ground limestone. Dissolve 1 lb of hydrated lime in 100 gal of water (1 g/liter or 1 tsp/gal), and stir thoroughly. Allow the mixture to settle and apply the clear liquid to the growing medium as a normal watering. The increase in pH will be ½ to 1 unit. In 2 weeks, test again. If still too low, reapply at the same rate. The effect is not long-lasting, but will be enough to raise the pH while the slower-acting ground limestone has a chance to react. Hydrated limestone is very caustic and can cause burns. Do not handle with wet hands; use eye protection. Sulfur reacts slowly to lower pH. If a rapid decrease is desired, irrigate with a solution of iron or aluminum sulfate.

When the plants are in the landscape, some control of pH can be achieved by selecting fertilizers with either acid or alkaline reactions. Fertilizer labels show the potential acidity or alkalinity of the product. Potential acidity is expressed as the calcium carbonate equivalent to neutralize a ton of fertilizer. The higher the potential acidity, the more acidifying the fertilizer will be. If the pH is too high, use an acidifying fertilizer; if low, a less acidifying fertilizer. Using fertilizers that minimize pH changes is best, but with proper selection, they can be used to offset pH changes that occur over time. The changes are usually small (< 0.5 pH unit) and occur over many weeks.

TISSUE TESTING

Another method of monitoring and managing the nutrition of foliage plants is analysis of plant tissue. Whereas analysis of the growing medium indicates the quantity of minerals retained within the substrate, foliar analysis tells one the quantity of nutrients actually absorbed by the plant. Ideally, foliage and growing medium samples should be submitted together so that the results can be compared and the best possible nutritional program developed.

As with soil testing, kits may be purchased from foliage testing laboratories. Directions for sampling are very specific and will be provided with the kit. Results and recommendations will be sent by the testing lab following analysis.

PLANTING

Plants may be installed in in-ground or aboveground planters either by direct planting or with the plants remaining in the growing pot.

Direct Planting. When installing plants directly in fixed planting beds in the ground, it is advisable to excavate the soil to a depth of 3 to 4 ft. Install a 4-in. perforated PVC pipe at the bottom of the bed, one line per 4 ft of width. Connect the outlet to the building's drainage system, cover with 10 to 12 in. of gravel, add a separator, and install the planting using a well-drained medium. Manual watering is recommended with hose bibbs located directly in the planter or not more than 30 to 50 ft away.

For direct planting in aboveground planters, provision must be made for excess water. Small to moderately sized planters may have drainage holes in the bottom, and saucers may be used to collect surplus water. Larger containers may have fixed drains in the bottom, or a vertical PVC pipe installed along the side permits excess water to be extracted with a syringe or portable sump pump. In both instances, several inches of ¾-in. gravel is placed in the bottom of the container, then a screen or fiberglass mat is installed to prevent the growing medium from infiltrating the gravel, and the planting is installed in a well-drained medium. Finally, a layer of mulch is applied (Figure 7.7).

Plants in Growing Pots. For ease of replacement, installing plants in their growing pots is desirable. When plants are used in groups, this practice facilitates removal and replacement of plants without injuring the root systems of others. Pots are simply lifted out and new ones put in their place.

The simplest method of installing plants using this system is to insert the growing pot into a jardiniere which is slightly larger. A mulch is usually used to hide the inner container.

In larger planters for either single plants or groups of plants, a layer of gravel plus the soil separator is installed. The planting is installed using peat, perlite, sand, or Styrofoam™ to fill the space between the containers. If pots vary in size, wooden blocks, bricks, or inverted pots may be used to raise them to the proper height. A mulch completes the installation (Figure 7.8).

Mulches. The surface of the planter is usually covered with mulch (Figure 7.9). Not only is water conserved and soil compaction and washing reduced, but it is decorative. The mulch is attractive and covers and hides the often-unsightly growing

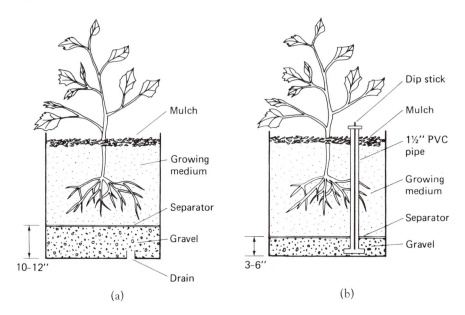

Figure 7.7 Planting directly into the growing medium: (a) fixed drain; (b) PVC pipe.

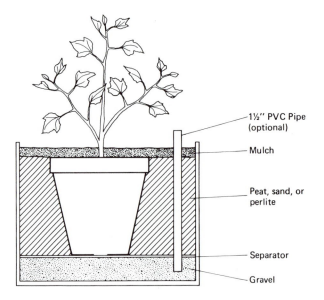

Figure 7.8 Double-potting, a method of installing plants kept in the growing container.

Figure 7.9 Mulching conserves water, reduces substrate compaction and washing, and is decorative.

container and medium. (Figure 7.10). If there is a large space between the inside and outside planters, the amount of mulching material required may be reduced, and maintenance facilitated, by placing a screen or other barrier on the top of the inside pot and then the mulch, rather than filling the entire space. (Figure 7.11). The space may also be filled with a lightweight, inert, waterproof material, such as Styrofoam pellets, which is covered with a barrier before applying the mulch.

Figure 7.10 Poorly installed plant. Mulch is needed.

Satisfactory mulching materials are sphagnum peat, bark, and stone. Organic mulches may be considered combustible decorative materials in some municipalities, and must be treated with an approved and registered flame-retardant before use. Interiorscapers should be familiar with, and comply with, the laws that apply at the site.

REPOTTING

Although plant growth is not a primary objective in interior plantscaping, the plants will produce some new stem and leaf tissue, and the root system will expand, if the environment is favorable. Eventually, repotting will be necessary.

Although it may be desirable for the plant to be slightly potbound, severe constriction of the roots is not beneficial. If the plants require frequent watering, and the roots are growing through the drainage holes of the container, they may be pot-bound. Determine the need for repotting by removing the plant from the growing pot and examining the root system. Invert small plants, and tap the pot sharply on the edge of a table to dislodge the plant. If the pot does not slide off easily, it should either be cut open or broken so as not to injure the roots. Larger plants may be removed by tapping the pot. A mass of matted roots twining around the soil ball indicates the need for repotting (Figure 7.12).

Never move the plant to an unnecessarily large container. The next larger size is recommended: a plant in a 5-in. pot is moved to a 6-in. pot, a 6-in. pot to a 7-in. pot, and so on. The procedure is a simple one, as shown in Figure 7.13.

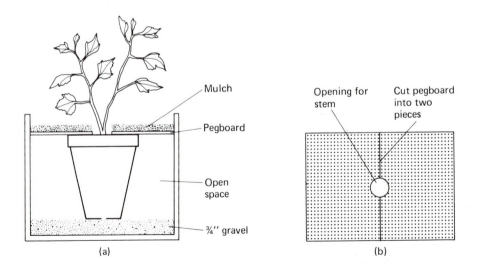

Figure 7.11 Method of conserving mulch (a), with detail of pegboard insert (b).

Figure 7.12   Plant in need of repotting.

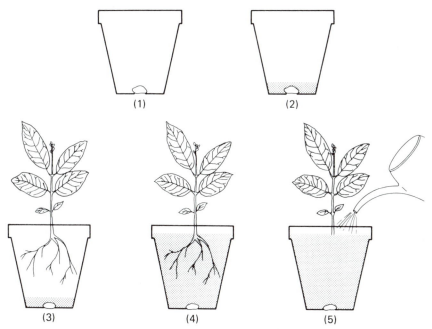

Figure 7.13   Reporting a plant: (1) cover the drainage hole; (2) add growing medium; (3) break mat of roots and insert plant; (4) add additional medium, allowing space at the top for water; (5) water immediately.

1. Do not use a layer of gravel, as this reduced the depth of the soil column and may impede drainage.
2. Add growing medium to the bottom.
3. Break the mat of roots and insert the plant.
4. Add medium to the sides; firm lightly, but do not pack.
5. Water immediately.

Care must be taken when repotting. A space at least equivalent to $\frac{1}{10}$ the depth of the container should be left between the medium surface and the rim to allow for water. Also, the plant should not be set any deeper than it was in the previous container, as the air relationships of the roots may be disturbed, causing injury. Following repotting, it is desirable to keep the plant in a reduced-light, high-humidity area for a few days until it recovers from the shock of the procedure.

In many instances indoors, the size of the container cannot be increased. In these situations, remove the pot-bound plant and break up the peripheral mat of roots. Carefully prune the roots. Replace the plant in the container, add new growing medium, and water thoroughly. Because a portion of the root system has been removed, thus reducing the ability to absorb water, it is desirable to curtail water loss by also reducing the evaporating surface. Pruning the top of the plant will restore the balance between roots and shoots. The repotted plants should be kept in a cool, shady, humid location for several days to reduce transpiration (evaporation of water) while new roots are produced.

SHRINKAGE

In a container, shrinkage refers to the reduction in volume of the medium, and usually results in increased water retention and decreased aeration. It occurs because of a disparity in particle size distribution and physical/biological factors acting on the individual components of the medium, especially microbial decomposition of organic components.

When media composed of different-sized particles are mixed, the final volume is less than the sum of the individual volumes of the components because fine particles settle into the large spaces between coarse particles. The greater the particle-size disparity, the more the shrinkage. To maintain, or restore, the correct volume, additional medium should be added periodically. To minimize potential injury to the root system, remove the plant from the pot, add the new medium to the bottom of the planter, and replace the plant.

HYDROPONIC CULTURE

In recent years interest has grown in raising plants without a growing medium. Plants are grown exclusively in water or in a nondecomposable material such as stone or other pellets. Water is readily available, and nutrients in the proper balance are provided either in solution or in time-released granules. Hydroculture is described in more detail in Chapter 9.

SUMMARY

The growing medium provides water and minerals and anchors the plant. A suitable medium must be porous and well drained, yet retain moisture and fertilizer elements. It should be standardized, uniform, inexpensive, and pest- and disease-free.

Natural soil is seldom used alone but may be mixed with amendments to provide desired physical characteristics. Organic materials such as sphagnum peat, sawdust, bark, or compost may be added, as well as coarse inorganic aggregates in the form of sand, vermiculite, perlite, rockwool, or calcined clay. Soilless media formulated from combinations of sand, peat, vermiculite, perlite, and bark are well suited for interior plantscapes. Pasteurization of all media is essential to eliminate pests, diseases, and weed seeds.

Media in containers hold more water than similar soils of greater depth. The number of large pores must be enhanced to facilitate adequate drainage and aeration.

Soil testing will facilitate evaluation of the fertilizer program and the diagnosis of plant problems.

Pot-bound plants should be identified and repotted. Care must be exercised to do the job properly. Shrinkage reduces medium volume and must be corrected to prevent overwatering.

REFERENCES

AIRHART, D. L., N. J. NATARELLA, and F. A. POKORNY: "Influence of Initial Moisture Content on the Wettability of a Milled Pine Bark Medium," *HortScience* 13(4): 432–434, 1978.

ATKINS, P. S.: "For Peat's Sake, It's an Excellent Medium," *Florists' Review* 172(4449): 19–21, 1983.

BAKER, K. F., ed.: *The U.C. System for Producing Healthy Container-Grown Plants*, Manual 23, University of California Division of Agricultural Sciences, Agricultural Experiment Station-Extension Service, Berkeley, Calif., 1957.

BEISEL, C. L., and D. L. HENSLEY: "Drainage Systems for Interior Plantings," *Interior Landscape Industry* 1(3): 47–48, 1984.

BLOM, T. J.: "Working with Soilless Mixes," *Florists' Review* 173(4480): 29–30, 32–34, 1983.

BOODLEY, J. W.: "Charting One's Way through the Maze of Soil Mixes," *Florists' Review* 169(4383): 20–21, 1981.

BOTACCHI, A. C.: "Superslurpers'-Water Release Crystals or Hydrogels," *Connecticut Greenhouse Newsletter* 151:21, 1989.

BUGBEE, G. J., and C. R. FRINK: *Quality of Potting Soils*, Bulletin 812, The Connecticut Agricultural Experiment Station, New Haven, Conn., 1983.

CONOVER, C. A., and R. T. POOLE: "Characteristics of Selected Peat," *Florida Foliage Grower* 14(7): 1–5, 1977.

____: "Composted Household Waste Utilized as a Medium Component in Greenhouse Foliage Plant Production," *Foliage Digest* 15(5): 5–8, 1992.

____: "Influence of pH on Activity of Viterra 2 and Effects on Growth and Shelf Life of *Maranta* and *Pilea*," *Foliage Digest* 3(1): 12–13, 1980.

DAWSON, M. D.: "Cation Exchange Capacities in Container Mixtures," *Foliage Digest* 3(11): 15–16, 1980.

DONAHUE, R., and J. MILLER: *An Introduction to Soils and Plant Growth*, 5th ed., Prentice-Hall, Inc., Englewood Cliffs, N.J., 1983.

DUNHAM, C. W.: "Nutrition of Greenhouse Crops in Soils with Added Peat Moss and Vermiculite," *Proceedings of the American Society for Horticultural Science* 90: 462–466, 1967.

EVANS, Y. E., I. SISTO, and D. C. BOWMAN: "The Effectiveness of Hydrogels in Container Plant Production is Reduced by Fertilizer Salts," *Foliage Digest* 13(3): 3–5, 1990.

FABER, W. R., and H. A. J. HOITINK: "Critical Properties of Successful Container Media," *Ohio Florists' Association Bulletin* 641: 2–5, 1983.

FOX, T. E. and E. R. EMINO: "Properties of Ground Cedar Mulch Amended Potting Mixes," *HortScience* 13(3): 349, 1978.

GARTNER, J. B.: "Amendments Can Improve Container Growing Media," *American Nurseryman* 153(3): 13ff., 1983.

HANAN, J. J.: "Oxygen and Carbon Dioxide Concentration in Greenhouse Soil-Air," *Proceedings of the American Society for Horticultural Science* 84: 648–652, 1964.

HARBAUGH, B. K.: "Root Medium Components and Fertilizer Effects on pH," *Foliage Digest* 17(3): 1–3, 1993.

HENLEY, R. W., and D. L. INGRAM: "Characteristics of Container Media Components," *Foliage Digest* 12(12): 1–8, 1989.

HERSHEY, D. R.: "Measuring Growing Media pH with Metal-probe and Glass Electrode pH Meters," *HortScience* 23(3): 625, 1988.

HOITINK, H. A. J., and H. A. POOLE: "Mass Production of Composted Tree Barks for Container Media," *Ohio Florists' Association Bulletin* 599: 3–4, 1979.

Horticultural Perlite for Successful Planting, Publication HP-77, Perlite Institute, Inc., 45 West 45th Street, New York, N.Y. 10036.

JOINER, J. N., ed.: *Foliage Plant Production*, Prentice-Hall, Inc., Englewood Cliffs, N.J., 1981.

_____, and C. A. CONOVER: "Comparative Properties of Shredded Pine Bark and Peat as Soil Amendments for Container-Grown *Pittosporum* at Different Nutrient Levels," *Proceedings of the American Society for Horticultural Science* 90: 447–453, 1967.

JUDD, R. W.: "All Peat Moss Isn't Equal; Be Careful in Selecting It," *Greenhouse Manager* 2(12): 74–75, 1984.

_____: "Peat Moss Compressed to Produce 'Hard Bale'," *Greenhouse Manager* 3(1): 89, 1984.

_____: "Soilless Mixes for Nursery Production," *Journal of Environmental Horticulture* 1(4): 106–109, 1983.

KOTHS, J. S., R. W. JUDD, JR., J. J. MAISANO, JR., G. F. GRIFFIN, J. W. BARTOK, and R. A. ASHLEY: *Nutrition of Greenhouse Crops*, Cooperative Extension Service, College of Agriculture and Natural Resources, The University of Connecticut, Storrs, Conn., 1976.

LOVE, J. W.: "Basics of Bark Media," *Ohio Florists' Association Bulletin* 584: 7–8, 1978.

MASTALERZ, J. W.: *The Greenhouse Environment*, John Wiley & Sons, Inc., New York, 1977, pp. 341–421.

_____: "Notes about Substrates," *Pennsylvania Flower Growers* 293: 6–12, 1976.

MOTT, R. C.: "Cornell Tropical Plant Mixes," *New York State Flower Industries Bulletin* 33: 5, 1973.

NATARELLA, N.: "Selecting Proper Soils and Fertilizers," *Florists' Review* 171(4424): 76, 78, 1982.

NEAL, K.: "A Look at Rockwool," *Greenhouse Manager* 8(5): 142–143, 1989.

Perlite, Technical Data Sheet 1-1 (1977), Perlite Institute, Inc., 45 West 45th Street, New York, N.Y. 10036.

PETERSON, J. C.: "Current Evaluation Ranges for The Ohio State Floral Crop Growing Medium Analysis Program" *Ohio Florists' Association Bulletin* 654: 7–8, 1984.

_____: "Effects of pH upon Nutrient Availability in a Commercial Soilless Root Medium Utilized for Floral Crop Production," *Ornamental Plants—1982, A Summary of Research*, Research Circular 253, Ohio Agricultural Research and Development Center, Wooster, Ohio.

_____: "Monitoring and Managing Fertility: II. Monitoring pH and Salt Levels in Growing Media," *Ohio Florists' Association Bulletin* 630: 3–4, 1982.

_____: "Monitoring and Managing Nutrition: IV. Foliar Analysis," *Ohio Florists' Association Bulletin* 632; 14–16, 1982.

POOLE, H. A.: "Costs and Losses of Soil and Soilless Media," *Ohio Florists' Association Bulletin* 584: 10–12, 1978.

POOLE, R. T.: "Changing the pH of a Potting Medium," *Foliage Digest* 8(7): 6–8, 1985.

____: "Potting Mixtures for Foliage Plants," *Interiorscape* 9(4): 54–55, 1990.

____, and C. A. CONOVER: "Light Weight Soil Mixes," *Foliage Digest* 2(6): 8–11, 1979.

____, and W. E. WATERS: "Potting Ingredients for Producing Quality Foliage Plants," *Foliage Digest* 4(1): 12–13, 1981.

SANDERSON, K. C.: "Growing with Artificial Media: The Advantages, Disadvantages," *Southern Florist and Nurseryman* 96(20): 13ff., 1983.

SHELDRAKE, R., JR.: "Artificial Mix Substrates in the U.S.A.," *Symposium on Substrates in Horticulture Other than Soils in Situ, ACTA Horticulturae* 99: 47–49, 1980.

____, and G. T. DOSS: "Is Your Mix Difficult to Wet?" *Florists' Review* 162(4191): 63–64, 1978.

SPOMER, L. A.: "The Basic Facts about Soils in Containers," *Florists' Review* 166(4314): 18–19, 60–62, 1980.

STEINKAMP, R.: "How to Take Media Samples from Container Grown Plants," *Florida Foliage* 18(9): 12–13, 1992.

"Sun's Rays Effective in Sterilizing Potting Soil," *PPGA News* 21(7): 8, 1990.

"Tracking Down the Proper Growing Medium," *Greenhouse Manager* 2(7): 55ff., 1983.

WHITCOMB, C. E.: "Drainage Factors in Plant Containers," *Florists' Review* 156(4051): 23, 56–57, 1975.

WHITE, J. W., and J. W. MASTALERZ: "Soil Moisture as Related to 'Container Capacity'," *Proceedings of the American Society for Horticultural Science* 89: 758–765, 1966.

Nutrition

Plants, like animals, require certain essential minerals for proper growth. In order to maintain an attractive interior garden with high-quality long-lived plants, one must understand and give careful attention to plant nutrition.

ESSENTIAL ELEMENTS

There are at least 16 elements required for plant growth. Carbon, hydrogen, and oxygen are major elements in the plant's organic structure and are usually readily available from air and water. The other elements are usually obtained from the growing medium, where the stock of nutrients is limited and some are rapidly depleted due to absorption by the roots and/or leaching. Fertilizer applications replenish their supply.

Nitrogen, phosphorus, potassium, calcium, magnesium, and sulfur are designated as macro, or major, nutrients because they are required by plants in relatively large amounts. No less important are micro, or trace, elements, which are required in very small quantities. Iron, zinc, manganese, boron, copper, molybdenum, chlorine, and nickel are the trace elements needed for healthy plants.

Nitrogen is a vital component of proteins, nucleic acids, chlorophyll, and many other cell substances; without it, plant life would be impossible. Phosphorus, too, is a component of every living cell, providing the energy for life's processes. Potassium is not a constituent of any known organic compound in plants, but is necessary as an enzyme activator for many cell processes. All 16 essential elements, in the proper balance, are indispensable to the normal activities of plants. Many of their specific functions are presently known, and these are summarized in Table 8.1.

TABLE 8.1   Roles of Essential Elements in Green Plants

Essential Element	Principal Role(s)
Nitrogen (N)	Constituent of proteins and nucleic acids (DNA, RNA), chlorophyll, phospholipids, some vitamins, all enzymes, many plant hormones, ATP and ADP, and numerous other compounds.
Phosphorus (P)	Constituent of all proteins, phospholipids, enzymes, sugar phosphates, nucleic acids, nucleotides, ATP, and ADP.
Potassium (K)	Involved in stomatal-control mechanism, maintaining electrical neutrality of cell, and enzyme systems of numerous processes. Influences the uptake of other minerals and their translocation in plants.
Calcium (Ca)	A constituent of the middle lamella. Affects membrane permeability and cell division. Activator of enzymes. Essential for nitrogen metabolism and normal protoplasmic functions.
Sulfur (S)	Part of certain proteins, some vitamins, and enzymes. Component of coenzyme-A used in respiration (Krebs cycle). Constituent of essential oils.
Magnesium (Mg)	Part of the chlorophyll molecule, activator of many enzymes. Associated with phosphorus metabolism and translocation and with maintaining integrity of ribosomes.
Iron (Fe)	Essential for chlorophyll synthesis. Constituent of many enzymes, including the cytochrome system (respiration).
Zinc (Zn)	Enzyme constituent, essential for the synthesis of indole-3-acetic acid (auxin).
Manganese (Mn)	Constituent of enzymes.
Boron (B)	Associated with water absorption and sugar translocation, and differentiation of meristematic cells.
Copper (Cu)	Constituent of enzymes.
Molybdenum (Mo)	Cofactor of enzymes in nitrate assimilation. Electron carrier in reduction-oxidation reactions.
Chlorine (Cl)	Enzyme activator.
Nickel (Ni)	Facilitates iron absorption.

A well-managed fertilizer program for indoor plants should provide adequate nutrition at all times. An insufficient or excessive amount of fertilizer in the medium is harmful to plants. Deficiencies which ultimately reduce and distort plant growth may be incipient for long periods before visual or gross symptoms become apparent. Once recognized and diagnosed, such insufficiencies may be easily corrected with proper application of fertilizer. Soil analyses at regular intervals will enable one to monitor the effectiveness of the fertilizer program and indicate any need for modifications.

Nitrogen is used by plants in relatively large quantities. In the nitrate form, it is readily leached (washed) from the soil and often becomes deficient. Symptoms of

Figure 8.1 (a) Nitrogen-deficient *Tolmiea;* (b) no deficiency.

nitrogen deficiency appear as a general yellowing or chlorosis of entire leaves, or the interveinal areas of leaves, beginning with the older ones and progressing toward the apex (Figure 8.1). Chlorophyll, the green pigment, disintegrates in the normal recycling process and, in the absence of nitrogen, new molecules cannot be synthesized. Chlorotic leaves eventually become dry, beginning at the tips; growth is slowed and stunted; and new leaves are small.

Phosphorus is easily locked insolubly in the growing medium in chemical combination with other elements and is practically nonleachable. Deficiency usually affects carbohydrate metabolism, resulting in a purpling or reddening of the veins on the undersurface of older leaves due to anthocyanin accumulation. Such coloration should not be confused with the normal red pigmentation found in *Zebrina* and *Coleus,* for example. Interveinal areas may appear abnormally dark green. Phosphorus-deficient plants may also be stunted with young leaves reduced to as little as one-third of normal size.

Deficiencies of potassium are not likely to occur if complete fertilizers are used. When natural soil is a component of the growing medium, the other major elements and the trace elements are not apt to present deficiency problems; however, toxicity can occur. Highly organic, soilless media require application of all elements. Special mixtures of trace elements are available for occasional use. In some species, for example *Gardenia* (Figure 8.2), lack of iron may cause interveinal chlorosis or yellowing of new leaves, which is readily corrected with an application of chelated iron to the growing medium. Iron deficiency is frequently associated with high soil pH, low soil temperature, or poor soil aeration. In addition to the iron application, proper steps should be taken to eliminate these causes. Typical deficiency symptoms are outlined in Table 8.2.

Excessive amounts of minerals in the growing medium may cause injury to plants, and generally death. Moderately high levels of nitrogen may lead to overabundant vegetative growth. In the low-light, high-temperature environments of many building interiors, this growth may be spindly and weak, resulting in unattractive plants. Leaching excess soluble nitrogen from the medium and adjusting the rate and frequency of nitrogen application should correct the problem. When large quan-

Figure 8.2   Iron deficiency in
Gardenia.

tities of fertilizer are used, soluble salts may accumulate, causing injury or death. This
subject is discussed more comprehensively in a later section of this chapter.

FERTILIZER

Fertilizer is a mixture of minerals applied to the medium to provide essential elements for plant growth. It is usually called "plant food," although this is a misnomer,
because plants manufacture their own food in the form of carbohydrates, fats, and
proteins. Organic and inorganic fertilizers are available, and either may be used.

Organic fertilizers are derived from organic materials such as dried blood,
bonemeal, cottonseed meal, fish emulsion, and sewage sludge. Minerals from such
sources are slowly available to plants because they are not water-soluble and must
first be acted upon by microorganisms before absorption by the roots. Organic fertilizers, although slowly available at first, release their minerals over a long period.
The rate of release is uneven, however, because temperature and composition of the
medium affect the degradation process. Organic fertilizers are generally safer to use
than inorganics because they are low in analysis and soluble salts are not likely to
accumulate in the growing medium, causing potential injury to plant root systems.
They are usually more expensive than other forms, and may improve the soil structure. Offensive odors from many of them may prohibit their use indoors.

Most inorganic, or chemical, fertilizers provide a source of minerals readily
available to plants. They are usually high in analysis, and, for equivalent nutrition,
a smaller amount is used compared with organics. Plants will respond equally well to
either type, provided that proper rates of application are used. For indoor gardens,
select fertilizer that will provide the needed elements most economically. Some
common fertilizer materials are listed in Table 8.3.

TABLE 8.2 Symptoms of Mineral Deficiency

Mineral	Deficiency Symptoms
Nitrogen	Stunting, interveinal, or general yellowing of leaves—oldest first. Leaves may die and fall from plant. Loss of vigor. Anthocyanin accumulation (reddening) in older leaves of some plants.
Phosphorus	Purpling and/or reddening of veins of older leaves, stunting, leaves reduced in size. Decrease in stem diameter. Leaves may be abnormally dark green.
Potassium	Chlorotic areas at leaf tips, along margins, and in interveinal zones, beginning on older leaves, may coalesce and become necrotic. Leaves may abscise. Reduction in plant size.
Calcium	Death of stem tips, poor root growth. Interveinal chlorosis of older leaves followed by necrotic lesions, which coalesce. Leaves turn brown and abscise.
Sulfur	General yellowing of younger leaves first. Leaves not dry. Petioles tend to be more vertical.
Magnesium	Chlorosis between veins of older leaves. No dead spots. May resemble potassium deficiency in later stages.
Iron	Interveinal areas of new growth yellow to yellow-green. Larger veins green. Leaves usually normal in size and undistorted in shape.
Zinc	Dwarfing, aberration of the root tips. Generalized leaf spots. Leaves reduced in width, giving narrow appearance, chlorosis.
Manganese	Similar to iron deficiency but with wider band of green along veins and veinlets. Interveinal area green-yellow. Scattered necrotic spots or streaks may develop. New leaves distorted.
Boron	Stunting. Immature leaves deformed. Terminal bud may die. Raised corky areas on underside of veins and along petioles, which ultimately turn black, sink, and exude gummy exudate. On vines, symptoms most pronounced at base of petiole. Short, thickened stems become tough and brittle. Chlorotic mottling and streaking (*Chrysalidocarpus*).
Copper	Terminal dieback with multiple budding immediately subtending terminals. New shoots soon die.
Molybdenum	Stunting. Leaves may have marginal scorching and cupping.
Chlorine	Bronze-colored necrosis of leaves. Plants prone to wilting.
Nickel	Interveinal areas of new growth yellow to yellow-green. Larger veins green.

Complete Fertilizers

Indoor plants are usually fertilized with a mixture composed of nitrogen, phosphorus, and potassium. This is a "complete" fertilizer—a misnomer, of course. Trace elements are supplied with special preparations if they are required.

The label of every fertilizer package shows the analysis, for example 15–15–15. The first number is the percentage of elemental nitrogen (N), the second indicates the percentage of phosphorus expressed as an oxide, phosphoric acid

TABLE 8.3   Some Common Fertilizer Materials and Their Analyses

Name of Material	Analysis	Long-term Effect on pH
Inorganics		
Ammonium sulfate, $(NH_4)_2 SO_4$	20–0–0	Very acid
Sodium nitrate, $NaNO_3$	15–0–0	Alkaline
Calcium nitrate, $Ca(NO_3)_2 \cdot 2H_2O$	15–0–0	Alkaline
Potassium nitrate, KNO_3	13–0–44	Neutral
Ammonium nitrate, $NH_4 NO_3$	33–0–0	Neutral
Monoammonium phosphate, $NH_4H_2 PO_4$	12–62–0	Acid
Diammonium phosphate, $(NH_4)_2 HPO_4$	21–53–0	Acid
Superphosphate, $Ca(H_2PO_4)_2 + CaSO_4$	0–20–0	Neutral
Treble superphosphate, $Ca(H_2PO_4)_2$	0–40–0	Neutral
Potassium chloride, KCl	0–0–60	Neutral
Complete		
Soluble		
Miracle Gro	15–30–15 + micros	Acid
Miracid	30–10–10 + micros	Acid
OHP General Purpose	20–20–20	Acid
Peters General Purpose	20–20–20	Acid
Peters House Plant Special	15–30–15	Acid
Peter Tropical Foliage	24–8–16	Acid
Slow-release		
Jobe's Plant Food Spikes	13–4–5	Acid
Mag-Amp	7–40–6	Acid
Miracle Gro Spikes	6–12–6 + micros	Acid
Once	16–8–12	Acid
Osmocote	14–14–14	Acid
Sierra	17–6–12 + micros	Acid
Liquid		
Jobe's Liquid for House Plants	1–2–1	Acid
Miracle Gro Liquid House Plant Food	8–7–6	Acid
Peters Liquid Plant Food	5–10–5	Acid
Schultz Liquid Plant Food	10–15–10	Acid
Organic		
Bonemeal	2–11–0	Acid
Fish emulsion	5–1–1	Acid
Liquid Kelp (seaweed)	None given	

(P_2O_5), and the third stands for the percentage of potassium oxide, potash (K_2O). Thus, a 15–15–15 fertilizer contains 15% N, 15% P_2O_5, and 15% K_2O by weight.

Because N and K are used more than other elements and are more readily leached than P is, fertilizers higher in these elements will yield best results. For foliage plants, fertilizers that contain equal percentages of NPK (1–1–1), or other ratios, such as 2–1–1 (twice as much N as P or K), 2–1–2, 3–2–2, or 3–1–2, are acceptable. Commercially available formulations with analyses such as 14–14–14, 20–20–20, 14–7–7, and 24–8–16, for example, conform to the recommended ratios and would be satisfactory. In soilless media, use of a fertilizer containing three parts nitrogen, one part phosphorus, and two parts potassium decreases soluble salts and reduces fertilizer costs.

Methods of Application

Some fertilizers are available in forms that must be applied dry as a top dressing or mixed with the medium during preparation. Others are water-soluble and are used as a substrate drench. The type is not critical as long as it provides the needed elements.

Except for slow-release, tablet forms, and superphosphate, dry fertilizers are not normally used with foliage plants. The time involved in measuring and applying small amounts of dry fertilizer to many individual containers usually makes this method expensive.

With a liquid-fertilizer program, one simply dissolves water-soluble chemicals at the correct rate and irrigates and fertilizes the medium at the same time. Thus, maximum control of the nutritional program is possible. There are numerous water-soluble materials on the market suitable for interior plantscapes. Fertilizer in liquid form ready for use is also available, but is usually much more expensive than that prepared from soluble powders at the time of use.

Whether fertilizer is applied in dry or in liquid form, it should be used at the appropriate rate. Infrequent small doses may not provide adequate nutrients, and high levels will cause injury. Fertilizers should always be applied to moist soil, and dry fertilizers should receive a thorough watering following their application to the growing medium. Hot water (180°F, 82°C) should be used to dissolve soluble fertilizers.

Improper placement of the correct rate of fertilizer may be harmful. Surface applied fertilizer placed next to the stem of a plant may cause localized lesions and girdle the stem. Fertilizers applied over the tops of plants may settle in the "vases" formed by the new growth and produce a notching or banded necrosis of young leaves. The effects of the improper fertilizer application are usually not noticed until several weeks later when the leaves have expanded.

The usual method of fertilizer application is to prepare the fertilizer solution in a watering can and apply a sufficient volume to thoroughly wet the substrate mass in each container from top to bottom. Where many plants are involved, this method might require numerous trips to the sink to mix additional solution and be time-consuming. In some commercial situations, use of a proportioner might be possible and speed fertilizer application.

A proportioner is a machine that mixes water and fertilizer from a concentrated stock solution to produce a final solution at the desired concentration. Where convenient to use, perhaps the simplest and least expensive proportioner is the Hozon. A Hozon siphons and mixes at the rate of about 1:12 to 1:22, depending upon water pressure. Before using a Hozon, one should determine the exact ratio by siphoning off 1 quart of water from the concentrate tank and measuring the quantity collected from the end of the hose. The total amount collected is the sum of the initial 1 quart siphoned plus the amount of water necessary to siphon up that quart. For example, if 16 quarts is collected, the Hozon is siphoning at 1:15. In actual use the concentrated fertilizer solution would be made 15 times as strong as it is to be applied to the growing medium. Suppose that fertilizer is to be applied at the rate of 1 lb/100 gal of water using a Hozon. Prepare a concentrated stock solution by dissolving 2.7 oz of fertilizer in 1 gal of water. Several fertilizer products contain dye to indicate fertilizer in the water. If no dye is present, some should be added. Attach the Hozon to the watering hose at the tap, drop the siphon hose into the concentrated stock, and turn on the water. Allow the clear water to run from the hose and begin fertilizing when the colored solution appears, indicating that the system is working. Dilute fertilizer will come from the hose at the desired rate of 1 lb/100 gal. Figure 8.3 illustrates the use of a Hozon.

Several types of small, portable proportioners containing stock tank and metering mechanisms are available and may be easier to use than a Hozon. Various types include M-P Mixer Proportioner, Dosatron, and Dosmatic Plus. Proportion

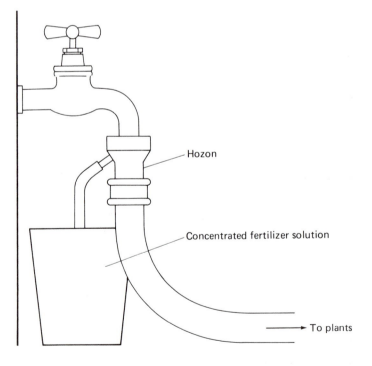

Figure 8.3   Use of a Hozon proportioner.

TABLE 8.4   Preparation of Fertilizer Stock Solutions
(oz Fertilizer/gal Stock)

Proportion Ratio	Rate of Application (lb/100 gal)				
	1	1¼	1½	2	2½
1 : 15	2.7	3.3	4.0	5.3	6.7
1 : 24	4.0	5.0	6.0	8.0	10.0
1 : 50	8.0	10.0	12.0	16.0	20.0
1 : 64	10.0	12.5	15.5	20.5	25.5
1 : 100	16.0	20.0	24.0	32.0	40.0
1 : 200	32.0	40.0	48.0	Not advisable	

rates vary from 1:50 to 1:500, depending on the equipment. Prices range from $75 to over $400.

In selecting a fertilizer proportioner, one should consider the initial cost, accuracy, mobility and portability, dilution rates required, capacity needed, and service available. They should be checked periodically for accuracy by using the method described above for a Hozon, or a conductivity meter, which measures ionic strength of a solution, may be used. Prepare a small sample of the desired concentration of fertilizer solution and determine its conductivity using the conductivity meter. Collect a sample delivered from the proportioner, determine the conductivity, and compare with the known sample. If the proportioner is not functioning properly, repairs should be made. Recommendations for preparing concentrated stock solutions for use with a proportioner are shown in Table 8.4. Figure 8.4 shows the use of the equipment.

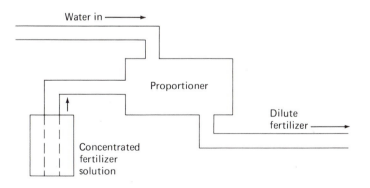

Figure 8.4   Use of a fertilizer proportioner.

Rates of Application

The major objective in interior plantscaping is to maintain plants rather than to have them grow; thus, the number of new leaves produced should be minimal. Since growth is less indoors, less fertilizer will be needed than for production.

The rate of fertilizer application is related to the light intensity and duration and is approximately one-tenth to one-fifth of the production rate. Plants maintained in high light for long periods require more fertilizer than do those maintained in less-than-optimal environments. Conover, Poole, and Henley have recommended yearly rates of nitrogen applications for several foliage plants at different light intensities (Table 8.5).

TABLE 8.5 Nitrogen Fertilization (g N/ft^2/yr) of Indoor Foliage Plants Under Various Light Intensities for 8 to 12 Hours Daily

| | Light Intensity (fc) | | | |
Plant	75–150	150–225	225–500	500–1000
Aglaonema (cultivars)	2	4	4	6
Araucaria heterophylla	2	2	4	4
Beaucarnea recurvata	2	2	4	4
Calathea (species and cultivars)	2	4	4	6
Chamaedorea elegans	2	4	4	6
Chamaedorea erumpens	2	4	4	6
Cissus rhombifolia	2	4	4	6
Codiaeum variegatum	2	4	4	4
Dieffenbachia (species and cultivars)	4	6	8	8
Dracaena deremensis (cultivars)	2	4	4	6
D. fragrans (cultivars)	2	4	4	6
D. marginata	2	4	4	6
Ficus benjamina (cultivars)	2	4	4	6
F. lyrata	2	4	4	6
Hedera helix (cultivars)	2	2	4	4
Howea forsterana	2	4	4	4
Maranta (species and cultivars)	2	4	4	6
Nephrolepis exaltata (cultivars)	2	2	4	4
Peperomia (species and cultivars)	2	2	4	4
Philodendron scandens 'oxycardium'	2	2	4	4
Philodendron (species and cultivars)	2	4	4	6
Radermachera sinica	2	4	4	6
Sansevieria (species and cultivars)	2	2	4	4
Schefflera arboricola	2	2	4	4
Spathiphyllum (cultivars)	2	4	4	6
Syngonium podophyllum	2	4	4	6

Source: Adapted from C. A. Conover, R. T. Poole, and R. W. Henley, "Light and Fertilizer Recommendations for the Interior Maintenance of Acclimatized Foliage Plants," *Foliage Digest* 14(11): 1–5, 1991.

The figures given in Table 8.5 are the base numbers used in calculating the amount of any fertilizer to apply. For example, suppose that you desire to fertilize *Schefflera* under 75 to 150 fc of light in a 10-in. pot using a 12–6–6 fertilizer. Determine the amount of fertilizer needed as follows:

1. Using the nitrogen portion of the fertilizer analysis, 12 in the example, determine the grams of fertilizer containing 1 g of nitrogen by dividing 100 by the nitrogen percentage:

$$\frac{100}{12} = 8.33 \text{ g}$$

2. Multiply grams of fertilizer by the base number for *Schefflera* under 75 to 150 fc, 2.

$$8.33 \text{ g} \times 2 = 16.66 \text{ g of } 12–6–6/\text{ft}^2/\text{yr}$$

3. See Table 8.6 to determine the number of pots of a specific size in a 1-ft^2 area.

4. Divide g/ft^2/yr by pots/ft^2 to determine fertilizer/pot. In our example,

$$\frac{16.66}{2} = 8.33 \text{ g of } 12–6–6/10\text{-in. pot}/\text{yr}$$

5. If divided into three applications per year, February, June, and September,

$$\frac{8.33}{3} = 2.78 \text{ g of } 12–6–6 /\text{application}$$

6. If 1 quart of fertilizer solution is applied to the medium, the fertilizer should be mixed at the rate of 2.78 g/quart or 11 g/gal. A small scale should be purchased to permit accurate weighing.

TABLE 8.6 Number of Pots of Various Sizes Required to Equal 1 ft^2 of Surface

Pot Size (in.)	Number of Pots/ft^2
3	16
4	9
5	7
6	5
8	3
10	2
12	1.3
14	1

Source: C. A. Conover and R. T. Poole, *Proceedings of the 1977 National Tropical Foliage Short Course,* p. 133.

Tables 8.7 and 8.8 show the amounts of a 20–20–20 soluble fertilizer and a 14–14–14 slow-release fertilizer, respectively, needed to supply the fertilizer levels suggested in Table 8.5 in various size pots. These recommendations should serve as guidelines and may require modifications depending on temperature (day and night), growing medium, source of light, and amount of water applied. Fertilizers containing high NH_4–N (ammonical–N, urea), such as 20–20–20, when used in peat-lite and pine bark media may be toxic. The microorganisms necessary to convert the ammonium to nitrate, while present in soil-based media, are not found in sufficient numbers in soilless substrates. With these media, high NO_3–N sources, including 15–15–15, 15–0–15, 15–16–17, calcium nitrate, and potassium nitrate are better. No more than half the nitrogen should be ammonical. Plants exposed to natural light require more fertilizer in the period

TABLE 8.7   Amounts of 20–20–20 Soluble Fertilizer Needed
to Supply Suggested Fertilizer Levels in Various Size Pots

	grams[a] 20–20–20/pot/4 months					
g N/ft²/yr	4"	6"	8"	10"	12"	14"
2	0.3	0.7	1.2	1.8	2.6	3.6
4	0.6	1.3	2.3	3.6	5.2	7.1
6	0.9	2.0	3.5	5.4	7.8	10.7
8	1.2	2.6	4.7	7.3	10.5	14.2

[a]One teaspoon of 20–20–20 equals approximately 5 grams.

Source: Adapted from C. A. Conover, R. T. Poole, and R. W. Henley, "Light and Fertilizer Recommendations for the Interior Maintenance of Acclimatized Foliage Plants," Foliage Digest 14(11): 1–5, 1991.

TABLE 8.8   Amounts of 14–14–14 Slow-Release Fertilizer
Needed to Supply Suggested Fertilizer Levels in Various
Size Pots

	grams[a] 14–14–14/pot/4 months					
g N/ft²/yr	4"	6"	8"	10"	12"	14"
2	0.4	0.9	1.6	2.6	3.7	5.1
4	0.8	1.9	3.3	5.2	7.5	10.2
6	1.2	2.8	4.9	7.8	11.2	15.3
8	1.6	3.7	6.6	10.4	14.9	20.3

[a]One teaspoon of 14–14–14 equals approximately 5 grams.

Source: Adapted from C. A. Conover, R. T. Poole, and R. W. Henley, "Light and Fertilizer Recommendations for the Interior Maintenance of Acclimatized Foliage Plants," Foliage Digest 14(11): 1–5, 1991.

from February to August, and little or none after that, as this is the time of year when growth is most active.

Although calculating the amount of fertilizer to apply using the procedure described above is the most desirable program, satisfactory results can be obtained by modifying the recommendations on package labels. Fertilizer at one-half to three-fourths of the recommended rate should be applied over a 3- to 6-month period. A single application may be made; however, more frequent applications at reduced rates, perhaps every 2 weeks for small pots to monthly for larger sizes, will maintain more uniform substrate nutrition, fostering maintenance of healthier plants. Simply apportion the fertilizer according to the number of applications involved. One-sixth the rate every 2 weeks will provide the same total concentration of nutrients as a full-strength treatment every 3 months. Knauss (1987) has suggested general fertilizer programs for interior plantscapes based on light intensity and nutritional needs (Table 8.9) of the plant. His recommendations are shown in Table 8.10.

TABLE 8.9 General Nutritional Needs of Some Common Interior Plants

Nutritional Needs	Plant
Low	Asparagus, Maranta, Peperomia, Sansevieria
Medium	Aglaonema, Araucaria, Calathea, Chamaedorea elegans, Cissus rhombifolia, Cordyline, Dieffenbachia, Dizygotheca, Dracaena (except D. marginata), Hedera, Hoya, Nephrolepis, Spathiphyllum, Yucca
High	Aphelandra, Brassaia, Chamaedorea erumpens, Chlorophytum, Chrysalidocarpus, Codiaeum, D. marginata, Epipremnum, Ficus, Philodendron, Syngonium, Polyscias

Source: Adapted from J. F. Knauss, "Thoughts for Food," Interior Landscape Industry 4(8): 39, 1987.

TABLE 8.10 Fertilization Suggestions for Interior Plants[a]

Light level	Frequency	Fertilizer Level (ppm Nitrogen)		
		Low	Medium	High to Very High
Low (100–300 fc)	1–2 times/yr	75[b]	150	225
Medium (300–1,000 fc)	2–4 times/yr	150	225	300
High (1,000–3,000 fc)	monthly during high-light months	150	300	375

[a]Based on Peters 24–8–16 water-soluble fertilizer.

[b]For Peters 24–8–18, 1/4 teaspoon per gallon equals 75 ppm nitrogen.

Source: Adapted from J. F. Knauss, "Thoughts for Food," Interior Landscape Industry 4(8): 39, 1987.

Frequent application at high rates, especially in the low-light and high-temperature environments of the average building, can result in soft, spindly growth and unsightly plants. Plant species, physical appearance, stage of development, size, season of the year, growing medium, and watering practices should all be considered when determining the time and frequency of fertilizer application. Generally speaking, most small, slow-growing, fine-rooted plants require less fertilizer than do large, rapidly growing, highly vegetative, coarse-rooted species. Soil test data are also useful for determining fertilizer needs.

When applying liquid fertilizer, a sufficient volume should be poured onto the moist medium so that some of it drains from the container bottom. This excess must be discarded. Such a practice assures application of sufficient fertilizer and its thorough distribution in the medium. If the plants are properly potted, filling the space between the medium surface and the rim of the pot should accomplish this objective. The slight excess will leach any residue from previous applications and reduce the possibility of soluble-salt accumulation in the medium and injury to roots.

Slow-Release Fertilizers

Fertilizers which provide nutrients slowly for a period of up to 6 months or more are available. Nutrients are encased in various types of coatings to form pellets which are mixed with the medium at potting or applied to the surface as a top dressing. Another method of application is to poke three or four holes, 1 to 2 in. deep, into the medium and fill with the appropriate amount of fertilizer. When water contacts the pellets, it penetrates the coating, gradually dissolving the minerals and making the nutrients available to the plants. For one, Osmocote, release of the minerals at moisture levels between the wilting point and container capacity is controlled by pH and microbial activity and is enhanced by increased temperature of the medium. Although slow-release fertilizers may be no better than liquid fertilizers, they do provide a constant source of nutrients for an extended period and save labor compared with the more frequent application of dry or liquid materials. The empty pellets may also improve soil structure, enhancing drainage and aeration.

Osmocote (14–14–14, 18–6–12) and Nutricote (14–14–14, Total 13–13–13, Total 18–6–8) are slow-release fertilizers available in quantity at reasonable cost. For a few plants, Osmocote is available in small retail packages, as are Once (16–8–12 + micros) and other slow-release fertilizers. Fertilizer tablets and spikes, when properly used, will also nourish plants satisfactorily for extended periods.

Micro Elements

For micro elements, the line between deficiency and toxicity is a fine one. Too much is worse than too little. Most soil-based media contain sufficient amounts of these elements for normal plant growth. Peat-based and bark mixes, on the other hand, contain small amounts of micronutrients, if any. Although some bark media may supply sufficient Fe, Zn, Cu, and Mn to satisfy a plant's needs, it is critical that micronutrients be added in well-balanced amounts either by incorporating them into the growing medium prior to planting or as part of the regular fertilizer program. Soil and tissue tests determine the need for micro-element fertilizers. Rates of application will be provided by the testing laboratory. In the absence of such data, follow the label recommendations.

Fertilizers are of several types. Some are inorganic compounds, such as iron sulfate, manganese sulfate, sodium borate, and others. Soluble forms are influenced by high pH and high phosphate and may be unavailable to plants. Chelated fertilizers, although more expensive, make iron, zinc, copper, and manganese, but not boron and molybdenum, readily available to plant roots even when the chemical environment of the medium is less satisfactory. Frits bind the elements into a glass base. They are expensive and require high application rates to be effective. Several commercial sources of micronutrient fertilizer are available, including Emisgram, Micromax, Peters Soluble Trace Element Mix, and OHP Minor Element Soluble.

Costs

Fertilizer cost is probably not an important consideration for the homeowner with a few plants. In maintaining extensive interior plantings, however, considerable savings may be made by comparing fertilizer costs. Nitrogen is the critical element in most fertilizer programs; thus, one should compare the cost of this element in materials otherwise acceptable. For example, if a 24–8–16 soluble fertilizer sells for $0.92 per pound, the nitrogen costs $0.24 per ounce, determined as follows:

1. One pound of 24–8–16 fertilizer gives 3.84 oz of nitrogen (16 oz × 0.24).
2. These 3.84 oz of N cost $0.92.
3. Cost per ounce = $\dfrac{0.92}{3.84}$ = $0.24.

A 10–15–10 liquid fertilizer for $3.24 per 12.0 oz has a nitrogen cost of $2.70 per ounce. The cost per ounce of all nutrients is $0.12 and $0.77, respectively, for the two fertilizers. If the costs of application are similar, the first is certainly a better value.

Nitrogen form, chemical purity, solubility, uniformity, effect on pH, method of application, ease of handing, and personnel expertise should be considered together with nutrient and application costs.

SOLUBLE SALTS

Soluble minerals in the growing medium are referred to as soluble salts. Such salts come from the parent minerals of the substrate, from applied fertilizer, and from the irrigation water. Low salt levels indicate that insufficient nutrients are available for absorption and the normal functioning of plants. High salt levels suppress growth or injure plant roots by causing reduced uptake of minerals and further water stress.

Figure 8.5 Toxic levels of salt injure or kill roots and prevent water and mineral absorption: (left) plant with a healthy root system; (right) root system exposed to high salts.

High levels of soluble salts inhibit water absorption by roots, and, in severe cases, water may be lost from plant roots to the growing medium. Toxic levels of salts injure or kill roots (Figure 8.5), totally preventing water and mineral absorption. Symptoms of soluble-salt injury of roots include wilting; chlorosis (yellowing of leaves); stunting; dying of the margins and tips of leaves, usually beginning on leaves near the stem apex and progressing toward the roots; leaf abscission; and injury or death of the roots. A crust of salts may also be evident on the surface of the growing medium.

Soluble salts may accumulate from any one or a combination of the following:

1. Application of fertilizer at an excessive rate.
2. Too-frequent fertilizer application.
3. Failure to apply sufficient water to thoroughly wet substrate mass, with some excess draining from medium.
4. Poor drainage of the medium.
5. Use of water that may be high in soluble salts, particularly calcium, magnesium, and sodium chlorides and carbonates.
6. Allowing plants forced in high-fertility programs to become too dry.
7. Use of soils with high fertilizer-retentive capacity.

Once recognized, immediate steps must be taken to reduce high soluble-salt levels. A thorough leaching (heavy watering) followed in 30 to 60 min by another leaching should remove much of the excess salt. Further buildup may be prevented by using sufficient water in regular waterings and by avoiding excessive amounts of fertilizer. Allowing hard water to sit for 24 hrs before use will permit many of the calcium, magnesium, and other ions to precipitate, reducing the potential for salt buildup.

If a high soluble-salt content is due to an excessive amount of slow-release fertilizer, removal is more difficult, as such materials cannot be leached easily. Heavy watering makes more nutrient salts available and may in fact compound the problem. If slow-release fertilizer is on the surface, it may be possible to remove the excess by inverting the container to allow the pellets to drop off. If mixed with the medium, one may be able to reduce the injury by carefully removing all the substrate from around the roots and repotting the plant in new growing medium.

Professional interior plantscapers may quickly and easily measure soluble-salt levels in the growing medium and irrigation water with a conductivity meter such as the Myron L Company Agri-meter, Figure 8.6, or a Solu bridge. Meters are available from greenhouse supply firms.

In using a conductivity meter, the soil sample is mixed with water, and after a certain waiting period, the solution is analyzed using the instrument. One method is as follows:

1. Sample the soil, avoiding the upper half inch. Allow to air dry.

2. Measure out a known volume of soil; for example, $\frac{1}{4}$ cup, and $\frac{1}{2}$ cup, and so on.

3. Mix the sample with two parts of distilled water. (Tap water may be used if a salt reading is made and subtracted from the reading for the soil–water solution.)

4. Stir the soil–water solution frequently for 45 min. For highly organic mixes, 3 to 6 hrs may be necessary.

5. Take the conductivity reading per manufacturer's instructions to determine salt content.

An interpretation of conductivity meter readings is shown in Table 8.11.

TOXIC SUBSTANCES

Fluorides.   Certain foliage plants, listed in Table 8.12, are susceptible to injury from fluorides, and may show brown necrotic (dead) spots or lesions, particularly along leaf margins or at the tips, especially in older leaves (Figure 8.7). Fluorides may come from fluoridated water, superphosphate fertilizers (1.5% fluorides), peat, vermiculite, or perlite added to the medium, and from the atmosphere. 'Baby Doll' *Cordyline* is injured by as little as 0.15 ppm fluorine.

TABLE 8.11   Interpretation of Conductivity Meter Readings
for a 1:2 Dilution (millimhos/cm)

Below 0.15	Very low, plants starved.
0.15–0.50	Satisfactory if medium is high in organic matter; too low if medium is low in organic matter.
0.50–1.80	Satisfactory for established plants.
1.80–2.25	Slightly higher than desirable.
2.25–3.40	Usually injurious.
Over 3.40	Excessive, severe injury.

Source: Adapted from J. C. Peterson, "Monitoring and Managing Fertility—Part II, *Ohio Florists' Association*, Bulletin 630: 3–4, 1982.

TABLE 8.12   Relative Sensitivity of Foliage Plant Species to Fluoride in the Growing Medium

Species	Common Name	Family	Sensitivity to Fluoride
Apidistra elatior	Cast-iron plant	Liliaceae	Suspected
Calathea spp.	Calathea	Marantaceae	Moderate–slight
C. insignis	Rattlesnake plant	Marantaceae	Moderate–slight
C. makoyana	Peacock plant	Marantaceae	Moderate–slight
Chamaedorea elegans	Parlor Palm	Palmae	Slight
C. seifrizii	Seifrizii palm	Palmae	Slight
Chlorophytum comosum	Spider plant	Liliaceae	Moderate
Chrysalidocarpus lutescens	Areca palm	Palmae	Moderate–slight
Cordyline terminalis 'Baby Doll'	Baby Doll Ti	Agavaceae	Severe
Ctenanthe sp. 'Dragon Tracks'	Dragon Tracks	Marantaceae	Slight
Dracaena deremensis 'Janet Craig'	Janet Craig	Agavaceae	Severe
D. deremensis 'Warneckii'	Warneckii	Agavaceae	Severe–moderate
D. fragrans 'Massangeana'	Corn plant	Agavaceae	Moderate–slight
D. marginata	Marginata	Agavaceae	Suspected
D. sanderana	Belgian evergreen	Agavaceae	Suspected
D. thalioides	Pleomele	Agavaceae	Suspected
Maranta leuconeura erythroneura	Red nerve	Marantaceae	Slight
M. leuconeura kerchoviana	Prayer plant	Marantaceae	Slight
Spathiphyllum spp.	Peace lily	Araceae	Slight
S. cannifolium	Peace lily	Araceae	Slight
Yucca elephantipes	Spineless yucca	Agavaceae	Moderate

Source: C. A. Conover and R. T. Poole, *Foliage Digest* 4(10): 5, 1981.

Figure 8.6 Agri-meter for rapidly measuring conductivity and pH. (Courtesy of Myron L Company.)

To minimize injury, the most important procedure is to maintain the pH of the medium above 6.0 to 6.5, thereby rendering most fluorides insoluble. Use dolomite, calcium carbonate, or hydrated lime to adjust the pH when preparing the medium. Drenching the acidic medium of potted plants with a suspension of hydrated lime at 1 to 2 lb/100 gal (1 to 2 tsp/gal) raises the pH and reduces fluoride availability.

The threat of toxicity may be further reduced by using water containing less than 0.10 ppm fluorine. Municipal water supplies may have as much as 1.0 ppm flu-

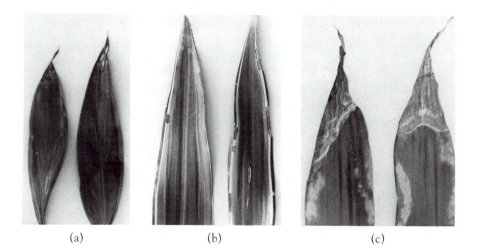

(a)   (b)   (c)

Figure 8.7 Fluoride injury: (a) *Cordyline*; (b) *D. deremensis* 'Warneckii'; (c) *D. deremensis* 'Janet Craig'. [From R. T. Poole, C. A. Conover, R. W. Henley, and A. J. Pate, *Foliage Digest* 1(7): 5, 1978.]

orine added to retard tooth decay and should not be used on susceptible species unless the water is allowed to sit at room temperature for 24 hours to permit dissipation of the chemical. Avoid using superphosphate and high fluoride amendments such as perlite and German peat in the potting medium. Leaching perlite before incorporating it with other components of the medium is a worthwhile practice for sensitive plants.

Chlorine.   Chlorine used to purify municipal drinking water is not present in sufficient amounts to cause plant injury. As with fluorine, aeration of water for 24 hrs will dissipate the chlorine.

Softened Water.   Cation-exchange water softeners frequently use table salt, sodium chloride, to replace calcium and magnesium and may produce water high in sodium. Toxic levels of sodium plus a possible increase in soluble salts may injure plants when softened water is used.

SUMMARY

At least 16 elements, mostly from the growing medium, are required by plants for growth, and many of their specific roles have been identified. Deficiencies or excesses of any of these minerals are expressed with typical symptoms and may be harmful or fatal to plants. Applications of organic or inorganic fertilizers at the correct frequency and rate replenish the stock of minerals used by plants and assure adequate nutrition. Excessive fertilization may cause high levels of soluble salts, resulting in injury or death if not corrected. Fluoridated water may be harmful, as may be water from cation-exchange softeners.

REFERENCES

ATKINS, P. S.: "Trace Elements Crucial to Growth," *Greenhouse Manager* 1(2): 78, 80, 1982.

BALL, V., ed.: *The Ball Red Book,* 14th ed., Reston Publishing Co., Inc., Reston, Va., 1984.

BLESSINGTON, T. M.: "Keeping Green Plants in the Pink," *Interior Landscape* 12(1): 32–38, 1995.

BOODLEY, J. W.: "Comparison of Four Trace Element Fertilizers in Peat-Lite Mixes," *Symposium on Substrates in Horticulture Other than Soils in Situ, Acta Horticulturae* 99: 33–38, 1980.

BOTACCHI, A. C.: "It Is Time to Change to Nitrate Nitrogen," *Connecticut Greenhouse Newsletter* 118: 1, 1981.

BROSCHAT, T. K.: "Nutrient Deficiency Symptoms in Container-Grown Plants," *Foliage Digest* 8(1): 1–4, 1985.

_____ and H. DONSELMAN: "Manganese Deficiency Symptoms in *Spathiphyllum*," *HortScience* 21(5): 1234–1235, 1986.

CONOVER, C. A.: "Effective and Economical Fertilizer Considerations," *Foliage Digest* 3(5): 8–10, 1980.

_____, and R. T. POOLE: "Basic Fertilization Guide for Acclimatized Foliage Plants," *Florists' Review* 168(4360):" 10ff., 1981.

_____: "Fertilization of Indoor Foliage Plants," *Proceedings of the 1977 National Tropical Foliage Short Course*, pp. 130–133.

_____: "Fluoride Analysis of Materials Commonly Available as Nutritional Soil Amendments," *Foliage Digest* 4(10): 5–6, 1981.

_____: "Fluoride Induced Leaf Mottling of *Dracaena Fragrans* 'Massangeana'," *Foliage Digest* 6(2): 6–8, 1983.

_____: "See the Light," *Greenhouse Grower* 1(7): 46, 48, 50, 1983.

_____ and R. W. HENLEY: "Light and Fertilizer Recommendations for the Interior Maintenance of Acclimatized Foliage Plants," *Foliage Digest* 14(11): 1–5, 1991.

Cooperative Extension Services of Northeast States: "Functions of Nutrients in Greenhouse Plant Production, Part 1," *Southern Florist and Nurseryman* 94(13): 26–27, 1981.

_____: "Functions of Nutrients in Greenhouse Plant Production, Part 3," *Southern Florist and Nurseryman* 94(15): 28ff., 1981.

_____: "Functions of Nutrients in Growing Greenhouse Crops, Part 2," *Southern Florist and Nurseryman* 94(14): 21–22, 1981.

DENEVE, B., and J. CIALONE: "Soluble Salt Level Control Essential to Plant Survival," *Florida Foliage* 10(9): 29–30, 1984.

DICKEY, R. D., and J. N. JOINER: "Identifying Elemental Deficiencies in Foliage Plants," *Florida Agricultural Experiment Stations, Annual Report 1958*, pp. 130–131.

DONSELMAN, H., and T. K. BROSCHAT: "Deficiency Dilemmas," *Interior Landscape Industry* 1(9): 30–35, 1984.

ELLIOTT, M., and D. B. MCCONNELL: "Anatomical Aspects of Fluoride Foliar Necrosis of *Cordyline*," *HortScience* 17(16): 912–914, 1982.

HARBAUGH, B. K., and G. J. WILFRET: "Correct Temperature Is the Key to Successful Use of Osmocote," *Florists' Review* 170(4403): 21–23, 1982.

HENLEY, R. W.: "Back to Basics—Soluble Salts," *Florida Foliage Grower* 12(3): 1–4, 1975.

_____: "Selecting a Fertilizer on More Than Price," *Southern Florist and Nurseryman* 94(46): 10, 12–13, 1982.

INGRAM, D. L., and R. HENLEY: "Determining Media Soluble Salts," *Foliage Digest* 6(2): 4–5, 1983.

JOINER, J. N., ed.: *Foliage Plant Production*, Prentice-Hall, Inc., Englewood Cliffs, N.J., 1981.

JUDD, R. W., JR.: "Adjust Fertilizers for Peat-Lite Mix," *Southern Florist and Nurseryman* 95(29): 26–27, 1982.

KNAUSS, J. F.: "Fertilizers for Injection," *Florida Foliage* 11(5): 22ff., 1985.

____: "Thought for Food," *Interior Landscape Industry* 4(8): 38–42, 1987.

KOTHS, J. S., B. GLEDHILL, R. W. JUDD, JR., J. J. MAISANO, G. F. GRIFFIN, J. W. BARTOK, and R. A. ASHLEY: *Nutrition of Greenhouse Crops*, Cooperative Extension Service, College of Agriculture and Natural Resources, University of Connecticut, Storrs, Conn., 1976.

MARLATT, R. B.: "Boron Deficiency and Toxicity Symptoms in *Ficus elastica* 'Decora' and *Chrysalidocarpus lutescens*," *HortScience* 13(4): 442–443, 1978.

____, and J. J. McRITCHIE: "Zinc Deficiency Symptoms of *Chrysalidocarpus lutescens*," *HortScience* 14(5): 620–621, 1979.

MAYNARD, D. N., and O. A. LORENZ: "Controlled-Release Fertilizers for Horticultural Crops," in *Horticultural Reviews*, Vol. I, Jules Janick, ed., AVI Publishing Company, Westport, Conn., 1979, pp. 79–140.

MOORMAN, G. W.: "Overfertilization," *Pennsylvania Flower Grower*, Bulletin 348: 4–5, 1983.

NIEMIERA, A. X.: "Micronutrient Supply from Pine Bark and Micronutrient Fertilizers," *HortScience* 27(3): 272, 1992.

PETERSON, J. C.: "Monitoring and Managing Fertility—Part II: Monitoring pH and Soluble Salt Levels in Growing Media," *Ohio Florists' Association* Bulletin 630: 3–4, 1982.

POOLE, R. T.: "Soluble Salts Interpretation," *Foliage Digest* 4(6): 11-13, 1981.

____, and C. A. CONOVER: "Boron and Fluoride Toxicity of Foliage Plants," *Nurserymen's Digest* 19(11): 92–94, 1985.

____: "Effects of Light Intensity and Fertilizer Formulation on Six Foliage Plants Growing Indoors," *Foliage Digest* 23(11): 7–8, 1990.

____: "Fluoride-Induced Necrosis of *Cordyline terminalis* Kunth 'Baby Doll' as Influenced by Medium and pH," *Journal of the American Society for Horticultural Science* 98(5): 447–448, 1973.

____: "Foliar Chlorosis of *Dracaena deremensis* Engler 'Warneckii' Cuttings Induced by Fluoride," *HortScience* 9(4): 378–379, 1974.

____: "Fluoride-Induced Necrosis of *Dracaena deremensis* Engler cv Janet Craig," *HortScience* 10(4): 376–377, 1975.

____: "Influence of Fertilizer, Dolomite, and Fluoride Levels on Foliar Necrosis of *Chamaedorea elegans* Mart," *HortScience* 16(2): 203–205, 1981.

____, R. W. HENLEY, and A. J. PATE: "Fluoride Toxicity of Foliage Plants—A Research Review," *Foliage Digest* 1(7): 3–6, 1978.

____, J. N. JOINER, C. A. CONOVER, and A. J. PATE: "Roles of Mineral Elements in Plants," *Foliage Digest* 2(2): 13–15, 1979.

"Researchers Find Nickel Essential to Plant Growth," *Interior Landscape* 9(6): 15, 1992.

SANDERSON, K. C.: "The Big N," *Florists' Review* 175(4538): 20–21, 1984.

____, W. C. MARTIN, JR., L. WATERHOUSE, and LIH-JYU SHU: "Slow Release Fertilizers Are Good for Houseplants," *Southern Florist and Nurseryman* 94(42): 29, 1982.

SHELDRAKE, R., Jr.: "Fluoride Toxicity: Dangers and Solution," *Florists' Review* 167(4341) 55–56, 1981.

____, G. E. DOSS, L. E. ST. JOHN, JR., and D. J. LISK: "Lime and Charcoal Amendments Reduce Fluoride Absorption by Plants Cultured in a Perlite-Peat Medium," *Journal of the American Society for Horticultural Science* 103(2): 268–270, 1978.

VLAMIS, J.: "Diagnosing Mineral Deficiencies and Toxicities in Plants," *Foliage Digest* 5(12): 8–10, 1982.

WHITE, J. W., and J. W. MASTALERZ: "Container Gardening Offers Something for Everyone," *Landscape for Living*, The Yearbook of Agriculture 1972, House Document No. 229, Washington, D.C.

Moisture

Plants cannot survive without water. Virtually all of the water used by a plant is absorbed by the roots from the growing medium, and an adequate supply must be available at all times. Excess water in the substrate inhibits aeration and may injure or kill roots. Overwatering is probably the most serious problem of plants in the indoor garden.

FUNCTIONS OF WATER

Water is essential for the growth and development of plants. It is the primary component of protoplasm, the medium in which all cellular reactions occur, a raw material in certain processes (photosynthesis, for example), and the medium for transport of materials within the plant. Water pressure causes turgor in herbaceous plants, holding them erect. In addition, cell enlargement, the dominant phase of growth, depends upon water. As vital as water is to a plant's existence, up to 95% of the water absorbed by the roots is translocated directly through the plant and evaporated from the leaves in the process of transpiration.

Aside from its direct role in plant growth and development, water may serve other functions. Irrigation fosters aeration of the growing medium, for as water fills the pores, old air is purged. With drainage, a suction is created which draws new air in from the top. Evaporation of water from the growing medium will increase the relative humidity of normally dry interiors, particularly in the area surrounding the plants. Some cooling of the air also occurs. Syringing plants with a stream of water may be useful in removing pests such as aphids and red spider, helping to reduce populations of these organisms. Syringing also removes dust and dirt, cleaning the leaves.

WATERING

Watering is the most repetitive and time-consuming task of the interior landscape technician. For conventional planters, it requires judgment as to whether or not water is needed and how much must be applied.

Frequency

Determining when to water is one of the more difficult aspects of maintaining indoor plants, but with practice, one can become quite adept. The growing medium should be watered when it needs it, not according to a predetermined schedule. If there is any error, it should be toward the side of dryness.

How often water is applied depends upon many factors, including the physical environment, age of the plant, kind of plant, size of plant, type and volume of container, growing medium, and plant growth activity.

The physical environment includes the light, temperature, humidity, and other factors surrounding the plant. Generally speaking, plants in bright light will require more frequent watering than those in low light, as they are more active physiologically. Plants in a cool room need less water than similar plants in a warm location, as warm temperatures and associated low relative humidity favor increased transpiration and water absorption, thus drying the growing medium. Not all plant species have similar water requirements. Cacti, for example, withstand a drier medium than do *Philodendron* or *Brassaia* and should be watered less frequently.

Large plants are usually grown in large planters containing a volume of growing medium which holds considerable amounts of water; therefore, they usually require less frequent watering than do small plants in small pots. Pot-bound plants of any size usually require water frequently.

The type of planter has a profound influence on watering frequency. Water evaporates through the walls of porous containers such as clay pots and plants in them must be watered more often than plants in ceramic, plastic, or other nonporous containers, the walls of which are impervious to water. Peat and bark media will not dry as rapidly as will a sandy medium of the same volume.

Plants require a constant supply of water to maintain their normal processes. During active growth they are watered more often than during periods of inactivity or when the environment is unfavorable. Old plants are usually not as physiologically active as younger plants of the same type, and may withstand more drying. Plants in bloom should be watered more frequently than nonbloomers.

The time of day one selects to water the plants is not important. There is a chance for disease to develop if the foliage remains wet for extended periods anytime either during the day or at night. Care in wetting only the growing medium and not the leaves will eliminate this problem.

Determining When to Water

The growing medium for foliage plants should never be allowed to dry out completely, and plants should never wilt. When a plant wilts, growth has already been reduced significantly. A single, severe incidence of water stress can cause serious loss of leaves in many indoor plants. Long-lived, quality plants cannot be maintained in the interior landscape if they are allowed to wilt as a routine practice. Before watering a wilted plant, be certain that lack of moisture is the cause of the problem. Wilting may also be due to overwatering or high levels of soluble salts which have injured the root system, curtailing water absorption. Low relative humidity accompanied by bright light and warm temperatures may foster a high rate of transpiration, and the roots, although in an adequately moist medium, are not able to absorb water as fast as it is lost. Nematodes and root rots may also cause wilting. Addition of water in these instances will not restore normal turgor to the plant and may aggravate the real problem. Proper diagnosis is important as are corrective measures. Improving drainage of the medium and the planter, increasing the depth of the planter, and reducing the frequency of watering will alleviate overwatering. Leach the soil to remove excess salts. Increase the humidity by syringing the plants, shading them, and lowering the temperature. Nematodes may be difficult to control even with proper chemicals, and the infested plant may have to be destroyed.

As the medium dries, there is an increase in soluble salts, with a 50% reduction in water content doubling the soluble salts and enhancing the possibility of root injury. Some media, such as peat-lite mixes, are difficult to wet once they are dry. Using a nonionic wetting agent according to directions on the label will speed the wetting of the medium.

Various methods are available to assess the water needs of the medium. Look at it. A wet medium will be darker in color than one that is dry. Feel it, stick your finger in the medium. If it is wet to the second knuckle of your index finger, moisture is adequate. In deep planters, probing 5 to 6 in. into the medium may be necessary to determine if water is needed. A clean trowel or bamboo cane inserted into a wet medium will meet resistance. Particles will adhere when the probe is withdrawn, similar to testing a cake with a toothpick (Figure 9.1). Probes which remove a narrow core of medium are also available. Where possible, lift the container. Wet growing media are heavy, and one will soon learn from the weight difference exactly when to water. In self-watering planters, the position of the water indicator float indicates the need for filling.

In recent years, moisture meters, such as that shown in Figure 9.2, have become available for assessing the water content of a medium. The instruments usually consist of a probe attached to a meter. Inserted into the growing medium, the probe conducts the electric current produced by the minerals in the substrate solution to the meter, where it is registered on a scale of moisture levels. When used

(a)   (b)

Figure 9.1   The need to irrigate may be determined by inserting a probe into the medium (a). If moist, particles will adhere to the probe when removed (b).

carefully, with readings properly interpreted, moisture meters can be a great aid in determining when to water. The soluble-salt level affects the reading, and if the probe is placed into distilled water, a moisture meter will read "dry."

Whether or not water should be applied must be determined for each plant individually, as they do not necessarily all need water at the same time. In fairly uniform interior environments, the watering frequency for each plant is quickly deter-

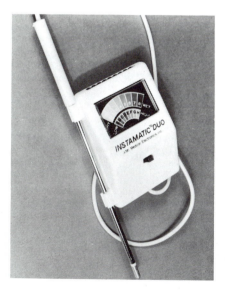

Figure 9.2   Moisture meter. (Courtesy of AMI Medical Electronics.)

mined and the watering process becomes routine. Plant maintenance firms usually service their contracts a minimum of once a week.

How Much Water Should Be Applied?

Apply a sufficient quantity of water to thoroughly wet the growing medium from top to bottom with some draining from the growing container. Small quantities of water applied at frequent intervals will not disperse evenly throughout the entire mass of growing medium. It wets to container capacity a depth of medium that varies with the volume of water applied. Thus, if small amounts of water are applied, the top layer is kept moist but the bottom remains dry. Because roots cannot grow in dry soil, the root system is confined to the narrow, moist zone near the surface of the medium and is further restricted in an already shallow planter with a small volume of substrate.

Applying a sufficient volume of water wets the entire growing medium and permits soluble salts to be leached. Water is applied again when the medium is dry. In planters with drainage holes, excess water should drain into the saucer and be discarded, thus preventing overwatering and reabsorption of leached salts.

In containers without drainage, care must be exercised so that excess water will not collect in the bottom. In a jardiniere, lift out the plant and dump out the excess water. In larger containers, the following procedure may be used:

1. In planting or double potting, install a 1- to 1½-in. polyvinyl pipe along the side of the decorative planter from top to bottom (Figure 7.7b).
2. Affix a "T" on the bottom of the pipe and cover with a plastic screen.
3. Cover the bottom of container and "T" with gravel.
4. Install the plants.
5. After each watering, use a dip stick made from a dowel or similar material to measure the water accumulation in the planter bottom.
6. Suck out any excess using a long-nosed syringe or portable pump.

WATER QUALITY

Water quality refers to the components in the water. When one considers all of the things that are either dissolved or suspended in water, it should be obvious that water quality can vary greatly from place to place. The three most important indicators of water quality are total soluble salts, hardness (alkalinity), and pH (acidity or basicity).

Soluble salts refers to all of the soluble minerals in the water. They are easily determined by measuring the water's electrical conductivity with a conductivity meter. The saltier the water, the more current it conducts. Readings from conductivity meters are usually expressed in millimhos, and should not exceed 1 millimho/centimeter. Multiplying millimhos/centimeter by 700 will yield a value that closely approximates parts-per-million soluble salts in the water.

Regular chlorinated tap water is suitable for foliage plants provided that it is neither high in fluoride nor softened. Some plant species are injured by fluoride, as described in Chapter 8, and fluoridated water should not be used to water susceptible plants.

Water from cation-exchange softeners is frequently high in sodium, which has replaced calcium and magnesium in the softening process. The soluble-salt level is increased and injury may occur. In addition, sodium ions destroy aggregates so that the structure of the medium may deteriorate from using softened water. Water used for plants should be drawn before it passes through the softener. Deionized water may be used satisfactorily.

Alkalinity refers to the water's capacity to neutralize acids and is generally known as hardness. High alkalinity is usually due to the presence of carbonates, especially those of calcium and magnesium, and the water will always have a high pH. All water with high pH does not have high alkalinity, however. Hard water can seriously interfere with the plant's ability to absorb minerals from the medium and may cause serious plant health problems.

The pH (acidity or basicity) of the water is easily measured with a pH meter. In general, water should be neutral to slightly acid for plant use. Water with high pH may have to be treated before using it on plants. Water from municipal systems is usually slightly basic and while not ideal for indoor plants, isn't as bad as hard water. The properties of water generally satisfactory for use on indoor plants are shown in Table 9.1.

Accidental spillage of commercial cleaning solvents onto the growing medium, or their contact with foliage, may injure some plant species.

Cold water may cause spots on leaves of *Philodendron, Syngonium, Sansevieria,* gesneriads, and other species (Figure 9.3) and tepid or room-temperature water should be used. Although cold irrigation water, 5 to 15°C, reduces the temperature of the growing medium, it returns to the preirrigation temperature in a short time.

TABLE 9.1   Properties of Water Generally Considered Satisfactory for Use on Indoor Plants

Electrical conductivity	<1 millimho/centimeter
Total soluble salts	<525 parts per million (ppm)
Percentage of salts as sodium	<40%
pH (acidity or basicity)	6.0–7.0
Hardness (alkalinity)	<100 ppm of carbonates
Sulfates	<240 ppm
Nitrates	<5 ppm

Source: Adapted from C. C. Powell, Jr., "Water Quality and Plant Health," *Interior Landscape Industry* 8(11): 45–46, 1991.

Figure 9.3   Cold-water injury on African violet.

In Florida tests, growth of *Aglaonema, Aphelandra, Dieffenbachia, Epipremnum, Maranta,* and *Nephrolepis* was not affected by cold irrigation water.

METHODS OF APPLYING WATER

Water is applied to the growing medium, not to the plants, either by overhead watering or by subirrigation.

Overhead Watering.   In overhead irrigation, water is poured onto the surface of the medium in a quantity large enough to wet the entire mass from top to bottom. If properly potted, filling the container to the rim will usually accomplish this objective. Apply the water with sufficient speed and force to accomplish the job swiftly and efficiently, but slowly enough to prevent washing of the medium, thus exposing roots to the possibility of injury or disease. Overhead watering may compact the medium, and a water breaker is recommended. For convenience, sinks with long spigot-to-basin distance are desirable. For floor planters, hose bibbs should be installed every 50 to 100 ft.

Water is usually carried to the site with watering cans, with either a 2-gal or 8-liter size being convenient and not difficult to handle. At home, smaller sizes may be more practical. Depending on accessibility of the water source, many trips to the tap may be necessary to water a large number of planters.

Portable equipment is available which provides running water wherever it is needed (Figure 9.4). Units consist of a water reservoir mounted on a cart and operated by a battery-powered pump or compressed air. Capacities range from 8.5 to 36

Figure 9.4 A portable watering system provides running water wherever it is needed.

gal. Purchased or made at home, these machines are filled at utility sinks or hose faucets. Even when full, the units are easily rolled to the plants where water is distributed through a flexible hose with nozzle. One manufacturer suggests that watering time may be reduced by up to one-half, thus cutting labor costs.

For extensive plantings, particularly those in the ground or for large containers, using a hose with a water breaker may be the best watering method. Metal or plastic extensions (Figure 9.5) will enable the technician to reach inaccessible planters. For home use, a mini-garden hose, available at garden shops, attaches to the kitchen faucet, and a 50-ft hose enables one to water plants throughout the house.

Trickle/Drip Irrigation. Designed initially for greenhouse and outdoor use, trickle, or drip, irrigation systems are being installed in interior plantings. Using inexpensive plastic components—pipes, tubes, and emitters—and operated manually or controlled by clocks, trickle irrigation systems precisely and constantly supply water to individual plants. Because water is delivered directly to the root system, there is little loss. Less water is used and conservation is enhanced.

A trickle irrigation system can be tailored to any need. Properly designed, such systems can be used for container plantings or display gardens. They are relatively inexpensive to operate, require small volumes of water at low pressure, and may be used to apply fertilizer.

If not properly installed, the growing medium may not be uniformly wetted. This happens because there is little lateral movement of water in soilless media. Tubes may also detract from the appearance of the planting and should be buried, if

Figure 9.5   Hose extenders facilitate irrigation of inaccessible planters.

possible. The water supply must be clear and clean to prevent clogging; filters are essential.

Irrigation design experts can help plan a trickle irrigation system. Equipment is available from greenhouse supply and irrigation supply dealers.

Subirrigation.   Subirrigation depends on the capacity of the medium to lift water from a reservoir and hold it against gravity. This capillary pull is a function of the small pores in the growing medium. The rise of water is inversely proportional to the size (diameter) of the pore; the smaller the pore, the greater the height to which water will rise in the medium. Capillarity is easily observed with a narrow straw in a glass of water, where the level of water in the straw will be higher than that in the glass. The narrower the straw, the higher the rise. In a well-structured medium, the large pores will drain freely and provide air for the roots, while the small pores will lift and hold water against gravity. Too many small pores result in overwatering of subirrigated plants.

Subirrigation (self-watering) planters work well and are being used with increasing frequency by both commercial interiorscapers and plant hobbyists. They can be used in new installations or to retrofit existing plantings. The systems available employ wicks or vacuum sensors.

Wick systems use either synthetic-fabric or soil wicks. Fabric-wick systems such as WaterDisc™, Figure 9.6, use cloth wicks similar to capillary matting. When installing a WaterDisc, carefully remove the plant from the grow pot and thread the wick through opposite drainage holes. Be sure the wick lies flat across the bottom and hangs evenly out each hole. Fit the fill tube through the lid of the water reservoir, attach the lid to the reservoir, and place the unit into a watertight, decorative container provided by the user. Place the empty grow pot with the wick on top of

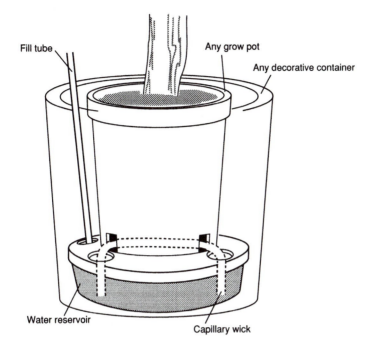

Figure 9.6 Wick system, WaterDisc.

the WaterDisc, insert the wick ends into the slots, and replace the plant. Cut the fill tube to the proper height and mulch as usual. Fill the reservoir and top water one time to prime the wick. Monitor the water level in the reservoir with a dip stick, and refill via the filling tube when needed. WaterDiscs are available in several sizes to meet a variety of needs.

Other cloth-wick systems, some of which use the decorative planter as the water reservoir, are available. Some have water gauges to indicate the need to refill. Small pots with watering wicks in saucers work on the same principle.

Soil-wick or soil-leg systems such as Everlife™, Jardinier™, and Mona Plant System™ (MPS), have a column of growing medium that extends from the medium into the reservoir. Water moves into.the growing medium by way of the soil wick. Users of Everlife and Jardinier have two options. They can purchase plants and planters separately and put them together, or they can purchase some plants from growers with the subirrigation equipment already in place. MPS is used in a manner similar to WaterDisc, except that a soil wick replaces the synthetic fabric. MPS units are available for use in individual planters as well as planter beds.

In the systems described, water moves constantly into the medium. Wetness is controlled by texture. Coarse-textured media with many large pores are not as wet as fine-textured media.

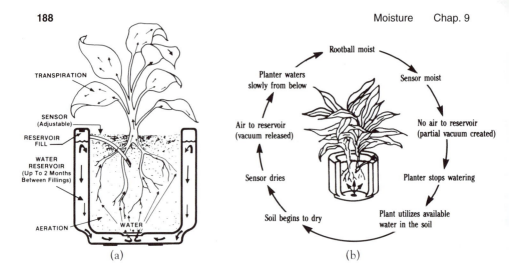

Figure 9.7 Vacuum-sensor system, Natural Spring: (a) components; (b) how it works. (Courtesy of Planter Technology.)

A vacuum-sensor system, Natural Spring™, uses a sensor to regulate the moisture content of the growing medium. Available in a variety of sizes, colors, and shapes, a double-walled planter sealed with a stopper is the water reservoir (Figure 9.7). Plants are installed by the user. Moisture flow is controlled by a porous sensor buried in the growing medium. As the moisture level of the medium drops, air penetrates the sensor, decreasing the vacuum in the reservoir. Water moves by capillarity upward from the base of the container through the growing medium.

When the sensor is moistened, air no longer enters the reservoir, stopping the water flow. When the sensor dries, the process is repeated. Thus a constant moisture supply is maintained. Raising the sensor in the potting mix increases water content, while lowering it reduces moisture. To work properly, the stopper must be in place in the fill hole.

Any watertight container is suitable for capillary watering. Several inches of coarse, clean gravel placed in the bottom serves as a water reservoir, with a plastic screen or weed matting on top of the gravel preventing infiltration by the medium. Install a fill column vertically and pot the plants in the preferred medium. Water initially from the top and then fill the reservoir periodically to maintain the water at the base of the medium.

Some tips on the management of self-watering planters are in order. Those using wicks or sensors require a sterile growing medium. Any well-drained soil-based or soilless medium that could be used successfully in a conventional pot is satisfactory. Hydroponic systems use clay pellets. All systems may be used to provide nutrients. The reservoirs may be filled with dilute nutrient solution, or pellets or packs of slow-release fertilizer may be used.

As water moves up through the medium by capillarity, minerals move with it. When the water evaporates from the surface, the salts are deposited and accumulate, forming a white crust. In time, the salts may become toxic. To leach systems containing a growing medium, water thoroughly from the top, collect the leachate in the water reservoir, and either pour or siphon it out. Leaching is accomplished in hydroponic systems by lifting out the planting basket and washing out the fertilizer. Other problems may occur when using subirrigation planters. Once defined, they are usually easily resolved.

Subirrigation systems often require significant up-front investment that may not yield dividends for some time. Interiorscapers should assess the available systems and weigh the potential benefits and drawbacks (see Chapter 6). Before making major changes, experiment on a small scale. Depending upon the nature of the plantings, one may have applications for more than one system.

At home, small pots may be watered from below by placing the containers on waterproof trays filled with several inches of moist sand, vermiculite, or perlite. To work properly, the growing medium must be in contact with the moist material in the tray, so drainage material is not used in the growing pot. Push the pots into the tray to assure contact. As additional water is required, it is poured onto the tray, not the surface of the medium in the individual containers.

Double-Potting. A modification of capillary watering is double-potting. The plant is planted in a recommended medium in a clay pot which is inserted into a slightly larger watertight container. The space between the inside and the outside pots is filled with sphagnum peat moss. Water is applied only to the peat and will be absorbed through the porous walls of the clay pot into the growing medium, satisfying the plant's needs.

Hydroculture. With hydroculture, the roots are constantly bathed in water. Used extensively in Europe, but still in its infancy in the United States, hydroculture is another method of watering which may be used successfully indoors. The plant is potted in a coarse aggregate of clay, shale, or coal. Plant roots grow into the pebbles as they would in soil. Water moves by capillarity through the pebbles (Figure 9.8). Moisture content varies from 100% at the bottom to less than 5% at the top, with the roots of each kind of plant seeking their own moisture level.

Subirrigation and trickle irrigation systems eliminate the time, effort, and tedium of manual watering. They do not replace human care in plant maintenance but give the technician more time for other meaningful maintenance, such as grooming and cleaning the plants. To achieve a long-lasting, high-quality interior plantscape, automated systems must be used in conjunction with the services of a competent horticultural specialist.

Spray Irrigation. These systems are designed to dispense volumes of water far in excess of those used by interior plants. Because they apply too much water too inaccurately, their use indoors is not practical.

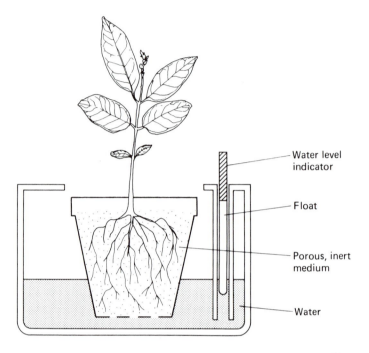

Figure 9.8 Hydroculture. A portion of the root system is constantly bathed in water. Water is added when the water-level indicator drops out of sight.

WETTING AGENTS

Technicians may spend a considerable amount of time trying to wet or rewet a growing medium that has become too dry. Large volumes of water are applied with little wetting of the medium. Most drains from the planter. This problem is associated with characteristics of water that affect how it moves in the growing medium.

Water molecules have high surface tension; they stick to each other rather than spreading out. Within the plant, such high surface tension facilitates translocation from roots to leaves. In the growing medium, it results in slow wetting, channeling of water, and runoff on and in the medium.

There are also adhesive tensions associated with water: how it clings to other things. Adhesive tensions cause water to perch at the bottom of the medium and not drain, and make it stick to the medium and be unavailable to the plant.

A group of materials has been developed which, when added to water, cause the water to penetrate more easily into, and spread more readily over, the surface of the growing medium. Called wetting agents, or surfactants, these materials reduce the surface tension of water. Water drops no longer cling together but spread out, forming a smooth, continuous film.

Wetting agents are a tool used by the interior plantscaper to facilitate water distribution and availability and to improve the drainage and aeration of the growing medium. They solve many water-related problems, including initial wetting, localized dry spots, rewetting, wilt, and compaction. Wetting agents are especially effective in soilless media, which are difficult to wet and rewet.

In selecting a wetting agent, one should consider toxicity, residual action, and cost. Nonionic forms should be chosen. Be sure that the material works with a variety of soil types and mixes. To minimize possible phytotoxicity, choose a product that is a blend of several wetting agents, not a single compound material.

The material used should be adsorbed (held in the medium) and give good results for several months, not one that must be applied with every irrigation. Cost is important. Available wetting agents range from 100% active ingredient to 90% dilute. Read the label and compute the cost per unit of active material.

Wetting agents are available as granules or liquids and may be incorporated into the growing medium or used in the first irrigation. In media containing bark and/or peat, they should be added when preparing the mix. For rewetting, apply every 4 to 6 months. Follow the manufacturer's recommendations for rates of application.

When properly used, wetting agents keep the growing medium in good, wettable condition. Because less water is wasted, their use reduces the amount of water applied and the frequency of irrigation.

ANTI-TRANSPIRANTS

Anti-transpirants are another group of chemicals that may be useful to the interior plantscaper. Sprayed on the leaves, they serve to reduce transpiration, thus protecting the plant from desiccation. Anti-transpirants may be used during transporting and transplanting to minimize "shock." In established plantings, they reduce water vapor loss associated with low relative humidity and air blasts from heating and air-conditioning systems. Intervals between irrigations may be increased, resulting in a saving of labor.

SUMMARY

Plants will not survive without water, and overwatering is a major plant problem. Plants should be watered when they need it—the frequency varying with the environment; the age, size, kind, and growth activity of the plant; and the type of planters and growing medium used.

The medium should never dry out completely. Sufficient water should be applied to thoroughly wet the entire mass of growing medium. Any excess should be discarded. Tepid tap water is suitable, and softened water should be avoided. Either overhead watering or subirrigation is satisfactory. Wetting agents facilitate wetting and rewetting and conserve water. Anti-transpirants reduce the drying associated with low relative humidity.

REFERENCES

BRESSAN, T.: "Drip Irrigation, The Basics," *Interscape* 5(36): 10–15, 1983.

CIALONE, J.: "The Underused Technology," *Interior Landscape* 10(2): 38–42, 1993.

DONAHUE, R., and J. MILLER: *An Introduction to Soils and Plant Growth*, 5th ed., Prentice-Hall, Inc., Englewood Cliffs, N.J., 1983.

DOSTAL, R. C., and J. W. WHITE: "Save Time and Money on Plantscape Maintenance," *Florists' Review* 164(4248): 20–21, 61–62, 1979.

GAINES, R. L.: *Interior Plantscaping*, Architectural Record Books, New York, 1977.

HIMES, L. B.: "To Water or Not to Water," *Plants and Gardens* 28(3): 15–16, 1972.

HYLAND, R.: "Interior Landscape Irrigation," *Interscape* 5(36): 6–8, 1983.

KAMP, M., E. ZUKAUCKAS, and M. SMITH: "Reducing Transpiration from Interior Plants," *Interior Landscape Industry* 1(9): 41–43, 1984.

MOORE, R. A., and D. M. POWELL: "Driving Your Soil to Drink- and Drain," *Florists' Review* 173(4480): 22–24, 26–27, 1983.

MOREY, J.: "New Trends for Interiorscapers," *Interiorscape* 3(3): 34–36ff., 1984.

OLSON, C.: "Planning for the Drip System," *Western Landscaping* 24(9): 19–23, 1984.

PEARSON, H. E.: "Effects of Waters of Different Qualities on Some Ornamental Plants," *Proceedings of the American Society for Horticultural Science* 53: 532–542, 1949.

POOLE, R. T., and C. A. CONOVER: "Growth Response of Foliage Plants to Night and Water Temperatures," *HortScience* 16(1): 81–82, 1981.

____: "Response of Foliage Plants to Industrial Cleaners," *Interiorscape* 11(4): 6, 39, 1992.

POWELL, C. C., JR.: "Water Quality and Plant Health," *Interior Landscape Industry* 8(11): 45–46, 1991.

REID, F. R., W. H. COWGILL, and A. W. CLOSE: "Plunging of Potted Plants in Relation to Moisture and Nutrient Supply," *Proceedings of the American Society for Horticultural Science* 45: 323–330, 1945.

STEVENSON, T.: "Plants in Containers Need Correct Watering," *The American Horticultural Magazine* 50(3): 118–120, 1971.

THOMPSON, L. M., and F. R. TROEH: *Soils and Soil Fertility*, McGraw-Hill Book Company, New York, 1977.

VERKADE, S. D., and D. F. HAMILTON: "The Benefits and Requirements of Trickle Irrigation," *Foliage Digest* 6 (9): 10–11, 1983.

WHITCOMB, C. E.: "Spare the Water and Save the Plant," *Horticulture* 54(1): 18–19, 1976.

WHITE, J. W., and J. W. MASTALERZ: "Container Gardening Offers Something for Everyone," *Landscape for Living*, The Yearbook of Agriculture 1972, House Document No. 229, Washington, D.C.

10

Problems

Various problems are encountered when growing plants indoors, some of which may cause a rapid decline in plant quality. These disorders may be caused by environmental and cultural conditions, pests, and diseases, or combinations of one or more factors. Plants should be examined regularly and treated immediately should any disorder occur. One should not expect plants to remain attractive indefinitely indoors. Even in the best environment plants will decline in time, and, when they become unsightly, should be replaced.

DIAGNOSIS

Diagnosis of the specific plant problem is essential for its correction, and is not always an easy task. Aphids are readily seen, for example, whereas Cyclamen mites are invisible to the naked eye. There are also many symptoms which look alike. Yellow leaves, for example, are symptomatic of many different plant problems, ranging from low light and overwatering to mineral deficiencies. Further complicating diagnosis is the fact that two or more problems may occur simultaneously.

To assist one in diagnosing plant disorders, a few tools are recommended (Figure 10.1):

1. Hand lens (10 to 20×).
2. Hand pH meter.
3. Conductivity meter.
4. Light meter.
5. Hygrometer.
6. Thermometer.

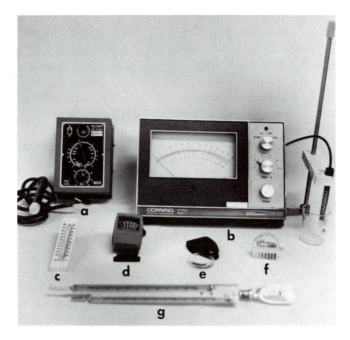

Figure 10.1   Equipment helpful in diagnosing plant disorders: (a) conductivity meter; (b) pH meter; (c) thermometer; (d) light meter; (e) hand lens; (f) pH indicator paper; (g) psychrometer (hygrometer).

The hand lens will enable one to examine tissue closely for the presence of pests and diseases. Important information about the mineral status of the medium may be quickly determined with the pH meter and the conductivity meter. The light meter will enable one to zero-in on problems related to light intensity, while the hygrometer will assess relative humidity, and the thermometer, the temperature. In addition, for closer examination of plant tissues, a dissecting microscope and a light microscope are valuable instruments.

Having the necessary tools, one should gather as much information as possible about the problem plant:

1. Cultural history.
2. Examine stems and leaves for pests and diseases.
3. Examine roots and root system.
4. Fluorine problem.
5. Environmental problem.
6. Human injury.
7. Pets.

Inquire about the cultural history of the plant. When was it placed in this location? How often is it watered? When was it last fertilized? What material was used and what was the rate of application? Has it been sprayed recently? With what? Have leaf-shining materials been used? What cleaning materials were used in the vicinity? Look at the container. What is its size and nature? Does it have drainage or not? To what extremes of temperature has the plant been exposed? Find out all you can .

Examine the leaves and stems thoroughly for pests and diseases. Pests may live in the axils of leaves or in the apex of the stem and be difficult to find, so look carefully. A hand lens is helpful. Diseases are not very common indoors because of the dry atmospheres, but they do occur. The presence of the mycelium (plant body) of a fungus or other symptoms may indicate disease, and diagnosis should be made.

If there has been little foliage wetting, look for a cause other than disease, as water is necessary for the growth and spread of the pathogen. If many different plants are affected all at once, this, too, is probably not disease, because disease organisms are fairly specific. One should suspect physical or environmental causes, even chemicals such as those used in floor cleaners or leaf shines.

Take soil samples for pH and soluble-salt tests. If possible, knock the plant out of the container. Look for evidence of disease. Are the roots white with many tips, or brown or black, indicating they are dead? Pull on some of the roots. Do they break cleanly or leave a fine thread behind? *Pythium*, *Phytophthora*, and *Rhizoctonia* are the most common root-infecting diseases. Look for root-infesting pests such as root mealybugs and fungus gnat larvae. Swellings on the roots may indicate the presence of nematodes. Examine the soil for slugs, snails, sow bugs, and other pests that may cause root injury. Be on the lookout for cultural mistakes. Is the plant potted too deeply? Is the soil compacted? Drainage poor? Medium too wet or too dry?

Fluoride in the water and fluorine in the atmosphere can injure plants. Is fluoridated water used? Is the pH correct? Was superphosphate used? Are there glass manufacturing or aluminum processing plants nearby?

Assess all parameters of the environment. What is the temperature? What are the extremes? Humidity? Light intensity? Is the plant subject to blasts of hot or cold air?

Physical injury frequently occurs to plants that are constantly brushed as people walk by. Have the plants been maliciously injured by human beings? Have pets been chewing on the plants? Has a cat been sharpening its claws?

Plants are exposed to a great many factors, and each one will exert a greater or lesser influence on plant growth and behavior. Not until one has obtained as much information as possible about a problem plant can a sound diagnosis of the disorder be made and control measures recommended.

PESTS

Numerous pests are known to attack foliage plants indoors. Any pest is to be considered a danger, as it may spread from one plant to another. Plants should be examined regularly, and any pests should be eradicated immediately.

Types

Aphids.   *Aphelandra, Brassaia, Gynura, Hoya, Dieffenbachia, Dizygotheca, Philodendron,* and *Saxifraga* are common hosts of aphids or plant lice, although many other foliage plants may also be infested. The insects are usually found on the new growth at the stem apex (Figure 10.2) or on the underside of young leaves, where they suck juices from the cells, causing deformed, curled growth of new leaves, buds, and flowers. Aphids excrete a sticky "honey-dew," which produces a high gloss on the leaf surfaces on which it accumulates. A black, "sooty mold" will grow on this honeydew, detracting from the plant's appearance. Aphids may also transmit disease.

The mature aphid is a small, soft-bodied, pear-shaped organisms $\frac{1}{16}$ to $\frac{1}{8}$ in. long. It is usually green in color, but may be yellow or black. Wingless forms are most common. Long legs and antennae are conspicuous. Reproduction may be very rapid, with a single female producing as many as 524,000,000,000 offspring in 8 weeks in a favorable environment. Female aphids have the ability of producing live young like themselves without mating. The insect molts as it grows, the shed skins providing an excellent diagnostic symptom. Winged forms may be produced and enable the insect to spread. Males may also be produced, and eggs laid following mating.

Mealybugs.   White, cottony patches covered with fuzzy, waxy threads, along leaves, veins, or in the axils, are an indication of the presence of mealybugs (Figure 10.3). Closely related to the scale insects, mealybugs suck plant sap, stunting plant growth and causing leaf wilting and abscission. Like aphids, they produce honey-dew.

The insects hatch from eggs deposited in a cottony, waxy sac produced by the female. The oval, light yellow, smooth-bodied young crawl over the plant and begin feeding by inserting their mouth parts into the plant tissue. A waxy material is exuded, covering the insect and making the plant unsightly. Mature insects are $\frac{1}{6}$ to $\frac{1}{4}$ in. long and may move about sluggishly. Winged males are produced which die soon after mating. The insects are persistent and difficult to eradicate, because chemicals do not readily penetrate the waxy covering.

Figure 10.2   Aphids, or plant lice, infesting new growth near the stem apex. (Courtesy of USDA.)

Figure 10.3   Infestation of mealybugs
along the stem and petioles of *Coleus*.
(Courtesy of USDA.)

Many foliage plants may be infested, including *Aglaonema*, *Anthurium*, *Aphe-landra*, *Araucaria*, *Ardisia*, *Asparagus*, *Asplenium*, cactus, *Cissus*, *Codiaeum*, *Coleus*, *Crassula*, *Dieffenbachia*, *Dizygotheca*, *Dracaena*, *Epipremnum*, *Ficus*, *Gynura*, *Hedera*, *Maranta*, *Nephrolepis*, palms, *Philodendron*, and *Syngonium*.

Red Spider Mites.   Mites are not insects, but Arachnids, and are closely related to spiders. The two-spotted, red spider mite is the most common mite on foliage plants, and the most prevalent pest indoors. The pest lives in the stem apex and on the underside of leaves, where it sucks plant juices. Mite-infested leaves turn silvery gray or yellow as a result of the destruction of chlorophyll. The margins of the leaf may die, and the leaf drop. Webs may also be present (Figure 10.4). In *Brassaia*, high levels of fertilizer favor high numbers of mites, and black areas may occur on foliage of mite-infested *Brassaia* moved from humid, low-light environments to dryer areas. In general, hot, dry conditions favor mites, with a single fertile female giving rise to 1,300,000 offspring in 30 days at 80°F. Mites readily travel from plant to plant on hands, air currents, clothing, and equipment such as feather dusters.

The adult female is a barely visible, eight-legged, reddish, green, or brownish organism about 1/60 in. long. Males are slightly smaller. Two dark-colored spots containing food show through the transparent body; hence the name. After mating, the female lays up to 70 spherical, shiny eggs on the underside of the leaf. Eggs hatch into small, crawling young. The female mite molts three times, the male twice, in the course of its growth. A generation takes up to 40 days. Since generations overlap, all stages are present at any time, making control more difficult.

Common hosts include *Anthurium*, *Aphelandra*, *Asparagus*, *Aspidistra*, *Brassaia*, *Codiaeum*, *Dieffenbachia*, *Dracaena*, *Fatsia*, *Hedera*, *Maranta*, palms, *Polyscias*, *Schefflera*, and *Syngonium*.

Scale.   About 20 species of scale are known to infest foliage plants. Soft brown scale is most common (Figure 10.5). *Agave*, *Aglaonema*, *Anthurium*, *Aphelandra*, *Araucaria*, *Asparagus*, *Aspidistra*, *Asplenium*, *Brassaia*, cactus, *Chlorophytum*, *Codiaeum*, *Dieffenbachia*, *Dracaena*, ferns, *Ficus benjamina*, palms, *Hedera*, *Nephrolepis*, and *Syngonium* are common hosts.

Figure 10.4   Red spider mites, the most common pest indoors, frequently spin webs. (Courtesy of USDA.)

Scale insects are $1/16$ to $1/8$ in. in diameter and are easily recognized by the shell that covers the body. Shapes vary from hemispheric, to oval, to oyster-like. Eggs, which hatch into nymphs or crawlers, are laid beneath the shell. The young move out from under the shell and crawl about for a short time. Finding a suitable spot, they insert their mouth parts into the cells of the stem or leaf and suck the sap. In a short time, a scale, or shell, forms over the body. Females remain under the shell. Winged males develop which move about actively and mate with the females.

Scale insects attack leaves and stems, sucking plant juices, causing stunting, leaf discoloration, and death of tissue. Honey-dew is also excreted. Because of the protective shell, control is very difficult, and almost impossible in the home. Since the crawler is the only stage not covered, controls are usually directed at them.

Whiteflies.   The plants, especially the undersides of leaves, are covered with small, white flies and oval, flat, pale green nymphs. They such the sap (Figure 10.6), reducing plant vigor and causing yellowing, death, and abscission of leaves. Honey-dew is excreted, which may become covered with sooty mold. Many foliage plants are attacked.

Eggs laid in clusters on the underside of leaves hatch into nymphs, which attach to the leaf and feed greedily. They grow and pass through four instars. In

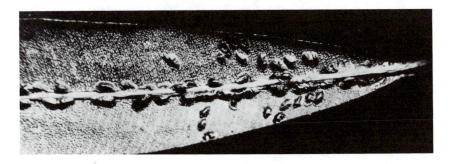

Figure 10.5   Soft brown scale on fern. (Courtesy of USDA.)

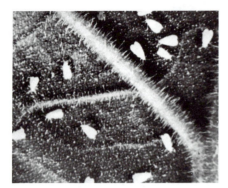

Figure 10.6 Whiteflies suck plant juices and reduce vigor. (Courtesy of USDA.)

about 30 days, adults appear as $\frac{1}{16}$-in., four-winged flies whose bodies appear as if dusted with a white material. Males and females fly and feed on the underside of leaves. Adults live 30 to 40 days. Generations overlap, and all stages may be found on infested plants at any time. Thus, control is difficult, requiring several treatments as frequent intervals.

Slugs. Slugs are mollusks, relatives of snails, clams, and oysters. They are gray to brown organisms $\frac{1}{2}$ to 4 in. long which feed on foliage, stems, and roots (Figure 10.7). A slime trail secreted by the organism will mark its presence. Slugs are night feeders and spend their days in damp places, such as under pots.

Eggs are laid in masses in damp places such as the growing medium. Within a month they hatch into small young resembling the adult except in size. They grow slowly and probably live for a year or more. Slugs may be collected and killed or controlled with baits or dusts.

Figure 10.7 Slugs are shell-less mollusks feeding at night on stems, leaves, and roots. (Courtesy of Foliage Education Research Foundation, Inc., Apopka, FL.)

Figure 10.8 Broad mite injury on *Aphelandra*. Newly formed leaves are deformed; shoot apex may die. (Courtesy of Foliage Education and Research Foundation, Inc., Apopka, FL.)

Cyclamen and Broad Mites. These mites (Figure 10.8) are pests, particularly in the Araliaceae—*Aralia, Dizygotheca, Fatsia, Fatshedera,* and *Hedera*—where they attack the young growth and buds, causing curled leaves, deformed buds, or death of the shoot apex. Webs are not produced. In *Hedera,* the mite causes production of stems either without leaves or with leaves that are small and deformed. The pest is small and transparent, and frequently is not detected.

Adults are shiny, oval, amber or tan, semitransparent organisms with eight legs. They are ¹⁄₅₀ in. long and may require a hand lens to be seen. The young are milky white and the eggs pearly white. All stages may be found on infested plants. A generation is completed in about 2 weeks.

Thrips. *Brassaia, Dracaena, Epipremnum, Fatshedera, Ficus, Nephrolepis,* palms, *Philodendron, Sansevieria,* and *Syngonium* are common hosts of thrips. These tiny insects rasp and shred the tissue of leaves and flowers. As the tissue dries around the punctures, the characteristic silvery, flecked appearance becomes evident (Figure 10.9). Distortion, yellowing, and abscission of leaves may occur. The underside of leaves becomes spotted with small black specks, the insect's excrement.

Eggs are deposited in slits in the leaf and hatch in 2 to 7 days into active nymphs. There are four instars before the adult emerges. Adults are barely visible, slender-bodied, tan, black, or brown in color, with lighter markings. A generation takes 25 to 35 days.

Fungus Gnats. Fungus gnats are one of the most common insects of interiorscapes. Plants in a medium infested with fungus gnat larvae lack vigor and have yellow leaves but show no apparent injury to stems or leaves. Roots show scars, or the small feeder roots have been eaten off by ¼-in. larvae or maggots which are embedded in the tissue or in the medium near the roots (Figure 10.10). Adults are tiny, black flies about ⅛ in. long, living near the substrate surface. Eggs laid in the

Figure 10.9 Thrips injury on *Ardisia*. Insects rasp and shred leaf tissue, which dries and produces the whitish- or silvery-flecked appearance symptomatic of thrips infestation. (Courtesy of Foliage Education and Research Foundation, Inc., Apopka, FL.)

medium hatch in 4 to 6 days into black-headed, transparent maggots which feed on root tissue for up to 2 weeks. The maggots pupate and emerge as adults in 5 to 6 days. Adults, which live for a week, mate and lay eggs for another generation. Adults fly at you when disturbed. They do not bite, but their presence is annoying. In addition, they carry and spread root diseases. Adults are most numerous during the establishment period of new plantings and control is most important at that time. Use of pasteurized growing media and biological controls will minimize fungus gnat populations.

Root Mealybugs.   Root tissue of *Asparagus*, bromeliads, *Chamaedorea*, *Chlorophytum*, *Chrysalidocarpus*, *Coffea*, *Cordyline*, *Dieffenbachia*, *Dizygotheca*, *Epipremnum*,

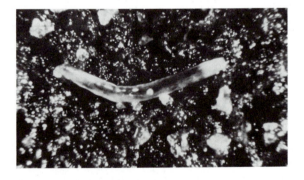

Figure 10.10 The ¼-in.-long larva of the fungus gnat feeds on plant roots, root hairs, and the lower stem. (Courtesy of Foliage Education and Research Foundation, Inc., Apopka, FL.)

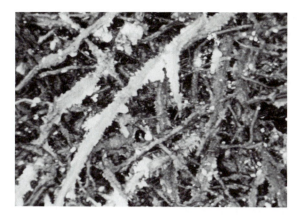

Figure 10.11 Cottony masses of mealybugs infest roots and root hairs, causing reduced plant vigor and foliar chlorosis. (Courtesy of Foliage Education and Research Foundation, Inc., Apopka, FL.)

Ficus, Hedera, Hoya, Peperomia, Philodendron, and *Syngonium* may be infested with the white cottony masses of root mealybugs. Reduced vigor, foliar chlorosis, and reduced growth may indicate root mealybugs (Figure 10.11). Strict sanitation is necessary to prevent infestation. Control is difficult, with pesticide drenches or granules the usual method.

Nematodes. Nematodes, or eelworms, are round worms which may attack foliage or roots of susceptible plants, sucking the juices (Figure 10.12). Leaves of *Anthurium, Asplenium, Begonia, Ficus, Peperomia, Pteris, Saintpaulia, Sinningia, Nephrolepis exaltata* 'Fluffy Ruffles,' and other species infested with nematodes may show symptoms similar to a bacterial or fungal leaf spot on the underside of leaves. The spots become brown and eventually black. The entire leaf may be affected. Adults may remain

Figure 10.12 Nematodes infesting the roots of foliage plants may produce numerous knots or galls. (Courtesy of USDA.)

alive in desiccated tissue for up to 3 years. Plant decline, foliar chlorosis, nutrient-deficiency symptoms, and wilting may also indicate the presence of nematodes.

In roots of Aglaonema, Anthurium, Asparagus, Brassaia, Caladium, Calathea, Chamaedorea, Codiaeum, Dieffenbachia, Dracaena, Epipremnum, Ficus, Maranta, Nephrolepis, Philodendron, Saintpaulia, Sansevieria, Schlumbergera, Siderasis, Syngonium, and other plants, nematodes may produce numerous lesions, knots, or galls. Stem galls in Mammillaria hexacantha produced by root-knot nematodes also occur. Other symptoms include swollen and injured root tips, dark lesions, excessive lateral branching, and stunted or restricted root growth. When present, there is a general decline in plant vigor. Nematodes are very small, $1/100$ to $1/8$ in. Worms hatch from eggs laid in galls and work their way through the medium to penetrate tender roots, and start a new infestation. Control is difficult. Use pasteurized growing media and good cultural practices. Infested plants have to be destroyed.

Control

Effective control of insects and mites indoors begins with identification of the pest and an understanding of its life cycle, since some pests have growth stages that are difficult to kill. Some cause injury only in one stage, others more. Control depends upon the stage present. In scale infestations, for example, crawlers are more easily controlled than are adults. Aphids, on the other hand, are easily killed. A good pest-control program indoors employs Integrated Pest Management (IPM) which involves manipulation of all plant management practices to control pests with a minimum use of energy and chemicals. IPM uses preventive (cultural), biological, and chemical controls. For IPM to be effective, early detection of a problem is imperative, and a continuous program of monitoring and scouting must be in place. Record-keeping is essential.

Preventive Control.   A preventive control program begins with the arrival of the new plants. Regardless of the program used in the production area, some pests are bound to survive on the plants you receive. New plants should be isolated from established stock and examined carefully. Any infestation, no matter how small, should be eradicated before the plants are offered for retail sale or installed in an indoor garden.

The practice of sound cultural techniques will assure the maintenance of quality plants. All cultural requirements should be as nearly optimal at all times as possible. A stressed or weakened plant is more susceptible to infestation. Good housekeeping is essential. Pick up dead leaves and flowers; destroy dead plants and those severely infested with pests or infected with disease. Don't use feather dusters or dust cloths unless they are rinsed between plants. Paper towels used once may be better.

If possible, pick off and destroy any pests. Vacuuming is also effective. Spraying the plants with tepid water will dislodge aphids, mites, and other pests. Washing the plant with mild soap and water and spraying with an insecticidal soap such as Safer's Soap are also effective pest controls. Be sure that both leaf surfaces are treated. In small infestations, mealybugs, aphids, and some other pests may be eradicated by dabbing the organism with alcohol on a cotton swab. Care must be taken not to touch the plant, as the alcohol may cause injury. Using resistant varieties, traps and baits, and sticky boards will also reduce pest problems.

Biological Control. Nonchemical methods of pest control are a possible alternative to traditional chemical means. Biological control may be successfully used in many interior plantscapes, particularly "closed systems," such as conservatory-type plantings, shopping malls, and building lobbies.

In biological control, natural enemies are introduced into the landscape to reduce insect and mite populations. The object is to keep the pest populations below the visible-injury level, not to eliminate them. Pests are reduced in several ways. Predators catch and consume their prey. Parasitoids typically lay an egg in or on its victim. The larva that hatches feeds on its host, ultimately killing it. Pathogens cause a disease harmful or lethal to the pest. Some commercially available biological control organisms are listed in Table 10.1.

TABLE 10.1 Some Commercially Available Biological Control Organisms (Predators and Parasitoids) for Use Against Insects and Mites in Interior Plantscapes

Pest	Biological Control (Common Name)	Biological Control (Scientific Name)
Aphids	Aphid Midge	*Aphidoletes aphidmyza*
	Green Lacewing	*Chrysoperla* sp.
	Aphid Parasite	*Aphidius matricariae*
	Convergent Lady Beetle	*Hippodamia convergens*
Mealybug	Australian Lady Beetle	*Cryptolaemus montrouzieri*
	Mealybug Parasite	*Leptomastix dactylopii*
Spider Mites	Predaceous Mites	*Phytoseiulus persimilis* (Figure 10.13)
		Phytoseiulus (Mesoseiulus) longipes
		Amblyseius (Neoseiulus) californicus
	Western Predatory Mite	*Metaseiulus occidentalis*
Soft Scale	Scale Parasites	*Macroterys flavus*
		Metaphycus helvolus
		Aphytis melinus
	Predaceous Lady Beetle	*Chilocorus nigritus*
	Scale Predator	*Lindorus lophanthae*
Thrips	Predaceous Mites	*Amblyseius cucumeris*
		Amblyseius barkeri
	Minute Pirate Bug	*Orius tristicolor*
	Thrips Parasite	*Thripobius semileutius*
Whitefly	Whitefly Parasite	*Encarsia formosa*
	Predaceous Beetle	*Delphastus pusillus*
Fungus Gnats	Predaceous Mites	*Hypoaspis (Geolaelaps) miles*
	Nematodes	*Steinernema* sp.
		Heterohabditis heliothidis

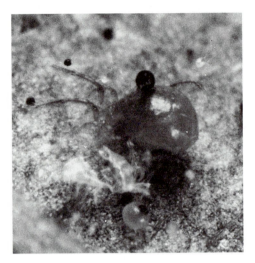

Figure 10.13   Predatory mite,
Phytoseiulus persimilis, consumes both
red spider mites and mite eggs.
(Courtesy of Richard K. Lindquist.)

Implement biological control programs at the first sign of infestation. For successful pest suppression, populations must be low when the beneficials are released. Pesticides may have to be used first to reduce pest numbers, as use of chemicals later may be harmful to the biologicals. Some predators will regenerate themselves, but the usual practice is regular release of new organisms. The number to use varies; your supplier can advise you. Some come in small vials and can be sprinkled onto the leaves; others arrive in a non-motile immature stage attached to small cards. All come with instructions for proper release. Beneficials should be released immediately upon arrival and not stored. Before release, be certain that they are healthy and thriving. Biological control organisms are available from commercial insectaries, greenhouse supply firms, and suppliers of organic garden products in the United States and Canada. Prices vary, so shop around.

Biological control offers many advantages. Client and public resistance to the use of chemicals is overcome, and without chemicals, the environment is safer and healthier. Plants are not harmed. In addition, costs may be cut compared with chemical control because no overtime is required, less time is needed, and no expensive chemicals must be purchased. No license is needed to distribute. The problem of pest resistance is eliminated. Beneficials are more specific than chemicals so they can be selected to deal with a specific pest. Finally, biological control is more effective than control by improperly applied chemicals.

There are certain disadvantages. Predators are expensive and must be released into the plantscape periodically to remain effective. They do not eliminate a pest, so that the client must be willing to accept some insects or mites. Additionally, some predators are active only at certain times, and since they are mobile, may not stay where you want them. Finally, biocontrols do not work as quickly as chemicals, so that the users, and their clients, must have patience.

Introducing IPM and biological controls is not an overnight process. Knowledge and patience are required. Although implementing the program can be expensive at first, long-term benefits will overcome the initial investment. IPM is often a process of trial and error and may not always be effective. When introducing a biological control program, begin on a small scale, an office for example, and expand as experience is gained.

Chemical Control. Chemical pesticides are poisons used to kill insects and other pests. Their use is difficult, if not impossible, indoors in most situations. They may also be harmful to human beings and animals, and should be used only as a last resort when a need exists for pest or disease control. Extreme care must be used when selecting and using chemical pesticides.

Selection. Before purchasing any pesticide, identify the problem and be certain that it can be corrected with pesticides. Read the entire label. The selected material must be one that will control the pest or disease successfully. It must not be toxic to human beings or animals and must not injure plants. Foliage plants are extremely susceptible to pesticide injury. The pesticide should not leave a conspicuous residue, nor should it have an offensive odor.

The sale, purchase, and use of any general- or restricted-use pesticide must be legal and conform to the Federal Insecticide, Fungicide and Rodenticide Act, as amended, and the related laws in each state. Businesses must be licensed and applications must be made by certified personnel. Licensing of businesses and applicators is the responsibility of each state.

Restricted-use pesticides include all products that bear a Poison–Danger statement and a skull and crossbones on the product label, except antimicrobial agents such as disinfectants, sanitizers, preservatives, and human and veterinary health products. There is an initial federal list of restricted-use pesticides published by the EPA in the *Federal Register,* Vol. 43, No. 28, February 9, 1978. Additions to that list are published periodically in the *Federal Register.* Each state may publish its own list of restricted-use pesticides. While a state list must include the federal list, it may be more extensive. Consult the Agricultural Extension Service in your county or your state Department of Agriculture for more information about certification or for a copy of the list of general- and restricted-use pesticides in your state labeled for use indoors on foliage plants.

Any pesticide must be handled and applied in accordance with the directions on the product label. The label is the letter of the law, and what the label says goes. If it can be used in an interior landscape, the pesticide must be so labeled. Labels on house plant pesticides may not say that the material can be used in the house, so read carefully. Few pesticides are registered for indoor use. Remember: Never do anything that is specifically prohibited on the label. It is against the law.

Beginning January 1, 1995, many pesticide labels began to list Worker Protection Standard (WPS) information. WPS is a safety measure designed to reduce pesticide-related illnesses and injuries to owners and managers of agricultural and horticultural businesses, their employees, pesticide applicators, and consultants. WPS applies mainly to people involved in production and research activities.

Interior plantscapers are not directly affected by WPS regulations unless they grow plants or treat stored plant material. A problem may be the restricted-entry interval during which entry into a treated area is virtually prohibited. One may have to wait to tag or select plants in field or greenhouse and experience delays in delivery of "add-on" orders. Interiorscapers should become familiar with the broad regulations of WPS. Time and trouble may be saved down the road. Information about WPS is available from the Department of Agriculture and the Agricultural Extension Service in each state and the Government Printing Office, Superintendent of Documents, P.O. Box 371954, Pittsburgh, PA 15250-1774.

Application methods.   The various methods for applying pesticides include sprays, aerosols, dips, drenches, systemics, and fumigants.

1. *Sprays.*   Applications of sprays may or may not be possible indoors. The presence of people or animals or potential damage to furniture, carpets, or drapes may prevent their use. Spraying during the night or on weekends when people are absent may eliminate the first obstacle.

Depending on the number of plants involved, one may use a hand atomizer or a small, compressed-air sprayer. An emulsifiable concentrate (EC or liquid) pesticide should be chosen in preference to a wettable powder (WP). Liquid formulations are easy to handle, remain in suspension, and seldom leave a residue on the foliage. The emulsifying agent is usually an oil, however, and may cause injury. Wettable powders contain no oils and are therefore less phytotoxic. They do not dissolve in water, so constant agitation of the spray tank is necessary to keep wettable powders in suspension. These materials usually leave a conspicuous residue which must be cleaned from the leaves and stems. Wettable powders should never be used as dusts.

Flowable pesticides are suspensions of finely ground wettable powders containing water and a wetting agent, thus enabling the user to mix and use them more easily. As with wettable powders, agitation during use is necessary, but, unlike wettable powders, flowables will not clog nozzles. Flowables may leave a worse residue on plant surfaces than wettable powders, however.

Add the concentrate or powder to the water and stir until thoroughly mixed. If the material does not spread well, add $1/2$ to 1 tsp of detergent per gallon or a spreader-sticker according to the label suggestions.

In applying a spray, one should be careful to cover thoroughly both leaf surfaces, stems, petioles, and axillary junctions. Spray to run-off. To minimize the possibility of plant injury, do not spray wilted plants or plants in full sun. Air and tissue temperatures above 90°F may increase chances of chemical injury. Do not spray when drying conditions are poor because leaves remain wet for long periods, and injury may occur. One of the most important ways to avoid plant injury is to make three to four preliminary sprays at 3- to 7-day intervals to a few plants. Injury will show up as soon as 18 hrs after application and certainly within 1 week.

2. *Aerosols.*   Aerosols are pesticide formulations in a pressurized "bomb." Many types are available for use on foliage plants. When using aerosols, follow the

label recommendations concerning the exact distance to hold the nozzle from the plant. When the gas is released from the can, it evaporates very quickly and may freeze close-by tissue. Most aerosol formulations will injure plants when applied at temperatures above 85°F or when the foliage is wet.

Before using any aerosol on foliage, make sure that it is recommended and safe for plants. Not all of them are. An ant and roach killer, for example, may contain oils which may be injurious to plants.

3. *Dips.* Pests may be effectively controlled by dipping the entire plant into the selected pesticide. Mix the pesticide in a container large enough to accommodate the top of the largest plant to be treated. Invert the plant and immerse it into the mixture for a few seconds. Protect the medium from falling out of the container with a plastic bag. Keep hands out of the dip unless you are wearing rubber gloves.

4. *Drenches.* Substrate-borne pests may be treated by drenching the growing medium with the correct pesticide mixture. Consult the label for correct rates of application. Do not apply soil drenches to plants under water stress or follow their application with additional water. Do not drench plants with pesticides meant to be used as a foliar spray as injury may occur.

5. *Granules.* Pesticide granules to be applied to the surface of the growing medium are now available. For safety, granular pesticides should be washed in with water, applied underneath mulch, or incorporated into the growing medium. Used according to the label, they will effectively control pests.

6. *Systemics.* The use of systemics in the interior environment is not recommended. Many of these materials are extremely toxic and could cause injury to human beings.

7. *Fumigation.* The nature of homes, retail shops, malls, and offices makes sealing these areas, and fumigation for pest control, impossible.

Using pesticides. Chemical pesticides are poisons. They should be used with extreme care and in strict accordance with the instructions on the label. Common sense should prevail at all times.

1. Read the entire label.
2. Wear protective clothing and use respiratory devices as listed on the label.
3. Avoid inhaling the mist.
4. Mix at the correct rate.
5. Mix only the amount needed; do not save leftover material.
6. Use only on the plants being attacked.
7. Apply only the amount needed.
8. Never smoke or eat while using.
9. Avoid contamination of food, dishes, and cooking utensils.
10. Keep people from the area, remove pets and their food and water, and cover aquaria, drinking fountains, floors, and furniture.

11. Clean equipment thoroughly.

12. Wash thoroughly after use and change clothing.

13. Store materials properly.

14. Do not save or reuse containers. Dispose of empty containers properly.

15. If you get the pesticide in your eyes, flush for 15 min with water and get medical attention.

16. If illness develops, call a physician; follow antidote instructions.

17. Maintain a file of Material Safety Data Sheets (MSDS).

Before purchasing any pesticide and prior to each use, read the entire label. Is it the right material for the problem to be corrected? What is the proper rate of application? What specific precautions must be followed? All instructions and precautions should be followed exactly.

Mix or dilute pesticides at your home base of operation. When handling pesticides, avoid contact of the chemical with the skin by wearing protective clothing. Arms and legs should be covered, and unlined rubber gloves should be worn. Wear rubber boots, as leather shoes and tennis shoes absorb pesticides. Mix the pesticides in a well-ventilated area, and when spraying, direct the nozzle away from oneself to avoid inhaling the mist. Respirators may be required for some materials, as indicated on the product label.

Mix the material at the correct rate as shown on the label. Using less than the recommended amount will not control the pest, and to do so would be a waste of time and money. Mixtures containing more than the appropriate amount of pesticide may be toxic to the plant, causing injury or death. Pesticide labels frequently specify rates per 100 gal. Table 10.2 shows the rates of application for small volumes of some frequently used pesticides.

One should mix only the amount needed. Should any pesticide be left over, it should be discarded properly, and less mixed for subsequent applications. If saved, the pesticide may corrode the spray equipment. In addition, it may deteriorate in storage, making it ineffective in destroying pests. Do not flush the surplus down the drain into the sewer or septic system, but dispose of it according to local regulations.

Treat only those plants being attacked, and apply only the amount needed. Sprays should be applied to both leaf surfaces to the point of run-off. Repeat applications may be required, with the interval between treatments depending on the pest. Proper timing is critical to successful pest control. Drenches should saturate the growing medium. For pots 4 in. or smaller, soil drenches are used at the rate of 1 pint dilute material/ft^2. For larger pots, 1½ to 2 pints/ft^2 is recommended. Never exceed 1 pint/ft^2 in mixes containing less than 50% peat. Wait at least 3 weeks before drenching again.

Care must be exercised so as not to contaminate food, dishes, and cooking utensils. Keep other people out of the area. Remove pets and their food and water. Cover aquaria. Do not smoke or eat while using pesticides. Never use your mouth to blow out clogged lines, nozzle tips, or other parts, or to siphon pesticides.

TABLE 10.2 Rates of Application for Non-Restricted-Use Pesticides Used in Small Volumes

Pesticide	Concentration/Gal Water
Banrot 15–25 WP	1 tsp
Captan 50 WP	3 tsp
Dexon 35 WP	3 tsp
Diazanon 4 EC	1 tsp
Diazanon 50 WP	1 Tbls
Kelthane 18.5 EC	2 tsp
Malathion 57 EC	2–3 tsp
Oil sprays	8–16 tsp
Orthene	1 tsp
Pyrethrum	Follow label
Resmethrin 2 EC	1 tsp
Rotenone 5 EC	4 tsp
Safer's Soap	9 tsp
Seven 50 WP	6 tsp
Terraclor 75 WP	1½ tsp
Truban 30 WP	1½ tsp

Thoroughly clean all equipment after use and store it where it will dry. Store the materials properly. They should be kept in closed, well-labeled, original containers in a locked cabinet in a dry, well-ventilated area free from temperature extremes. Keep them out of the reach of children and pets, and away from food or feed. Under the sink, in a pantry, or in the medicine cabinet are not suitable places to store pesticides. Do not save or reuse pesticide containers, but dispose of them in accordance with local regulations. Storage of pesticides for more than 2 years is not recommended, as deterioration of the product may occur if kept for longer periods.

After using pesticides, wash thoroughly, even take a shower, and change your clothes. If any material is spilled on either skin or clothing, wash and change clothes immediately. Clothing exposed to pesticides should be washed daily. Do not mix with other laundry. Pre-rinse before washing, then wash in hot water (140°F/60°C) containing the manufacturer's recommended rate of heavy-duty liquid detergent. Use normal or full water levels and a 12-min wash cycle. Wash 2 to 3 times. Line-dry the clothes to avoid contaminating the dryer. Rinse the machine thoroughly after laundering by running a complete cycle with detergent, but without clothes.

Any pesticide that gets into the eyes should be flushed out at once with water and medical attention secured. If illness develops during or after using a pesticide, call your physician or poison control center. Read the label or MSDS naming the active ingredients and follow all antidote instructions.

Control recommendations. Pest control recommendations are revised constantly. After identifying the specific problem, consult your local pesticide dealer or the county agricultural agent for the current control recommendations.

Figure 10.14   Plant injury may occur if a pesticide is phytotoxic to a particular plant, or when a recommended material is used improperly. Here, using Malathion at more than the recommended concentration has produced necrotic areas and caused abscission of leaves on *Ficus benjamina.*

Pesticide injury.   Plants may be injured if a pesticide has been used which is phytotoxic to the particular plant, or when a recommended material is used improperly (Figure 10.14). The new growth is most likely to be affected, resulting in stunting and abnormal growth such as twisted and distorted young leaves. Other symptoms include necrotic leaf tips or margins, chlorosis, depressions, scab, black leaf spots, and leaf burn either singly or in combinations. Injury is often evident in 18 to 72 hrs after treatment, and within 1 week if it is going to occur at all.

Injury may be reduced or eliminated by selecting the correct material and using it in strict accordance with the directions. By not applying pesticides to stressed plants, there is less danger of injury.

DISEASES

For a pathogenic disease to develop, three conditions must be satisfied. There must be a suitable host plant, a pathogen, and an environment that fosters growth of the pathogen.

The host may be any one of the several hundred species of foliage plants used indoors. Pathogens include various fungi, bacteria, viruses, mycoplasma-like organisms, and viroids capable of infecting from single plant species to a wide range of plants if environmental conditions are favorable. Most disease pathogens are favored by wet or humid conditions accompanied by warm to hot temperatures. Should any of the three requirements be absent, disease will not occur. The environment of the average building interior, with its low humidity, is not conducive to the onset of disease. Those that do occur are usually secondary infections following a nonpathological stress (excess water, for example). Eliminating the stress usually corrects the problem.

Common symptoms of diseased plants include stunting, chlorosis, leaf spots, leaf scorch, leaf abscission, and rotting of the roots and stems. Even though a disease is diagnosed and controlled, disease-damaged foliage will never return to normal.

Root Rots

Diseased roots are a common occurrence of foliage plants that have been improperly watered. The growth of several pathogens is favored by an excessively wet medium. Symptoms typically include wilting of the plant; yellowing and abscission of leaves, usually from the base up; brown to black root rot; and death of the roots. Examination of the root system will confirm the presence of pathogens.

Rhizoctonia attacks many plants, including *Aglaonema, Aphelandra, Ardisia, Begonia, Brassaia, Chamaedorea elegans, Cissus, Dieffenbachia, Epipremnum, Fatsia,* ferns, *Fittonia, Gynura, Hedera, Hoya, Maranta, Nephrolepis, Peperomia, Philodendron scandens* 'oxycardium,' *P. pertusum (Monstera deliciosa), P. selloum, Pilea, Pittosporum,* and *Syngonium.* All types of plant tissue may be infected, with diseased tissue producing a prominent reddish-brown mycelium, the body of the fungus. The medium surface, too, may be covered with the fungus.

Pythium is a water mold whose growth is favored by wet conditions. Many foliage plants are susceptible to infection by *Pythium,* including *Aglaonema, Aloe, Anthurium, Aphelandra, Araucaria, Ardisia, Begonia, Brassaia,* cactus, *Chamaedorea elegans,* Christmas cactus, *Chrysalidocarpus lutescens, Cissus, Dieffenbachia, Epipremnum, Fatsia, Fittonia, Hedera helix,* palms, *Peperomia, Philodendron scandens* 'oxycardium,' *P. bipennifolium, P. pertusum (Monstera deliciosa), P. selloum,* other *Philodendron, Pittosporum, Pilea, Spathiphyllum,* and *Syngonium.*

Infected plants usually wilt, often with a progressive yellowing of the leaves from the base up. The roots collapse and blacken, beginning at the tip and working back. In time, the exterior of the root will slough off, leaving the inner core (stele), giving a thread-like appearance to the root.

Phytophthora will infect *Aglaonema, Dieffenbachia, Hedera, Kalanchoe,* palm, *Philodendron scandens* 'oxycardium' and other *Philodendrons, Peperomia, Saintpaulia,* and other species. It is closely related to *Pythium,* and may attack foliage, causing leaf spots, and mature plant tissues, resulting in complete collapse of the plant.

Sclerotium is known to attack *Brassaia, Chamaedorea elegans, Dieffenbachia, Dracaena godseffiana, Epipremnum, Peperomia, Philodendron micans, P. scandens* 'oxycardium', *Pilea, Spathiphyllum, Syngonium,* and other foliage plants. Typical symptoms include a heavy, white fungus growth on the substrate surface and affected plant. White to tan sclerotia (spore-like resting bodies) about the size of mustard seeds are evident.

Fusarium has been shown to cause root rot in *Aglaonema, Asparagus,* cactus, *Dracaena,* and palms. *Cylindrocladium* has been shown to cause root and petiole rot in *Spathiphyllum.* Lower leaves turn yellow and sometimes wilt, with the entire plant eventually becoming chlorotic. Roots are blackened and mushy. Petioles may rot off. Root rot of *Araucaria* may also be associated with *Cylindrocladium*.

Prevention is the best defense against root-rot organisms. The growing medium should be well aerated, easily drained, and have been pasteurized. Care should be taken not to overwater. If disease develops, reduce the frequency of watering, and perhaps repot. A fungicide may be drenched onto the substrate to help control the pathogen. No one material will kill all the species that may be involved, so that proper identification is imperative before a fungicide is applied. The fungicide is mixed according to the label directions and should be applied at the rate of 1 pint/ft^2 for pots less than 4 in. in diameter, and at $1\frac{1}{2}$ to 2 pints/ft^2 for larger pots. If the mix contains less than 50% peat, never exceed 1 pint/ft^2. A repeat drench should not be applied for at least 3 months.

Stem and Leaf Diseases

Stem and leaf diseases, while not common on foliage plants indoors, do occur. Sound cultural practices should be used at all times in order to minimize the possibility of disease infection. Particular attention should be given to proper watering so as not to overwater the plants or to excessively wet the foliage. Crowding of the plants should be avoided. Destroying diseased leaves and/or plants will prevent further spread. With all foliar pathogens, except powdery mildew, keeping the foliage dry will control the pathogen without the need of chemical application. Appendix B is an index of some stem and foliar diseases found on foliage plants. Once diagnosed, a recommended fungicide or bacteriacide will eradicate the disease.

CULTURAL AND ENVIRONMENTAL PROBLEMS

Most of the problems that one encounters on foliage plants indoors are not caused by pests or diseases, but are due to plant stress caused by cultural or environmental factors. All aspects of the environment interact simultaneously to modify the effects of the individual parameters, with the plants themselves being the best indicators of cultural conditions. When cultural conditions are proper, plants develop deep green foliage, compact stems, good root systems, and well-expanded, new leaves. When out of balance, disorders develop. Many of these problems, their causes and corrections, have been discussed in previous sections of this book.

To review, cultural and environmental problems may relate to:

1. Growing medium.
2. Light.
3. Temperature.
4. Moisture (Figure 10.15).

Figure 10.15 Overwatered *Peperomia.*
Wilting is a typical symptom.

5. The atmosphere.

6. Physical injury (Figure 10.16).

Typical symptoms of various culturally and environmentally produced problems are shown in Table 10.3. There are many look-alikes and several problems may overlap. Proper diagnosis is important. Not all yellow leaves are equal, for example. Many factors will cause leaves to turn yellow, as shown in the table. Depending on

Figure 10.16 A burned margin occurred when this leaf rested against a hot-water heating pipe.

TABLE 10.3 Typical Plant Problems and Their Causes

Symptoms	Too Much Fertilizer	Too Little Fertilizer	High Salts	Fluoride	Dry Soil	Overwatering	Improper pH	Too Low Light	Too High Light	Light From One Side	Root Rot	Pot Bound	Too Hot	Too Cold	Cold Water	Low Humidity	High Humidity	Poor Soil Drainage	Pollution	Pesticide Injury	Physical Injury	Disease	Normal Aging	Drafts	Pests
No new growth	X													X		X									
Weak growth		X	X					X			X	X	X	X		X		X	X			X	X		X
Slow growth	X	X	X		X	X		X			X	X	X	X		X		X	X			X			X
Stunting		X	X		X	X	X	X			X	X	X	X		X				X		X			X
Small new leaves		X	X		X		X	X			X	X	X	X		X			X	X		X			X
Yellowing of old leaves		X	X		X	X	X	X	X		X		X	X		X			X	X		X	X		X
Yellowing of new leaves		X	X		X		X	X	X				X	X					X	X		X			X
Burned leaf tips	X		X	X	X				X				X							X		X			X
Burned leaf margins	X		X	X	X				X				X	X						X		X			X
Leaf spots			X			X			X				X	X	X	X				X		X		X	X
Leaf drop	X		X		X	X	X	X			X	X	X	X		X			X	X		X	X	X	X
Leaf curl									X				X	X		X				X		X		X	X
Wilting	X		X		X	X	X	X	X		X	X	X	X		X		X	X	X		X			X
Rots at soil line						X												X				X			
Plant bends toward light										X															
Stems broken, leaves torn																					X				
Death of roots	X		X		X	X					X		X	X				X	X			X			
Leaf color faded		X						X	X				X						X						X
Algae on surface of medium	X							X																	
Large, deep green new leaves	X							X																	
Leaves purplish								X						X											

216

whether yellowing begins with old or new leaves, or is general throughout the plant, will help in identifying and correcting a specific problem.

Once the cause of the problem has been determined, the necessary corrective measures may be taken. Consult the appropriate sections of the text for suggestions for correction of cultural and environmental problems.

SUMMARY

Problems affecting indoor plants may be caused by environmental and cultural factors, pests, diseases, or combinations. Diagnosis is not always easy, and one should gather as much information as possible about the problem plant before attempting a diagnosis.

Common pests include aphids, mealybugs, red spider mites, scales, and whiteflies, plus slugs, cyclamen mites, thrips, fungus gnats, root mealybugs, and nematodes. Pest control should employ Integrated Pest Management (IPM) and be preventive, because use of chemicals indoors is difficult. Eradicate the pest before moving the plant indoors, use sound cultural practices, pick off and destroy pests, and spray with tepid water. Biological controls are an important component of IPM programs.

Chemicals should be used only when a specific need exists. Proper materials must be used at the proper rates. Read the label and follow all instructions and precautions. Application is usually as sprays, dips, or aerosols.

Foliage diseases are not very common indoors because of the very dry air. When drainage of the medium is poor or overwatering is a problem, root rots may occur. Examination of the root system will confirm the presence of pathogens.

Cultural and environmental problems may be related to growing medium, light, temperature, moisture, the atmosphere, and physical injury, singly or in combination. Elimination of the stress will correct the problem.

REFERENCES

BURCH, D. G.: "The Question Approach to Diagnosing Plant Problems," *Interior Landscape Industry* 2(4): 28–36, 1985.

CARLSON, J.: "Be Aware of Your Pesticide's Shelf Life," *Southern Florist and Nurseryman* 94(1): 57–58, 1981.

CATHEY, H. M., and L. E. CAMPBELL: "Plant Distress Signals," *Interior Landscape Industry* 1(3): 29–34, 36–39, 1984.

CHASE, A. R.: "Common Fungal Root Rot Diseases of Foliage Plants," *Foliage Digest* 4(7): 15–16, 1981.

____: "Diagnosis of Foliage Plant Diseases," *Nurserymen's Digest* 18(7): 86–88, 1984.

____: "Indoor Plant Problems," *Foliage Digest* 17(7): 1–4, 1994.

____: "Phytotoxicity Concerns for Foliage Growers," *Interior Landscape Industry* 1(2): 78–80, 1984.

____: "Reducing Stress for Healthy Plants and Disease Control," *Foliage Digest* 7(3): 5–6, 1984.

____, D. T. KAPLAN and L. S. OSBORNE: "Nematode Pests of Tropical Foliage Plants and Leatherleaf Ferns," *Foliage Digest* 6(7): 3–6, 1983.

DAVIDSON, A. D., B. G. WESENBERG, and O. C. MALVY: *Maladies of Ornamental Plants*, Extension Bulletin 608, Cooperative Extension Service, Washington State University, Pullman, Wash., 1971.

DENMARK, H. A.: "Damage of Interior Foliage Plants by Banded Greenhouse Thrips," *Foliage Digest* 1(4): 8–10, 1978.

GEISTLINGER, L.: "Following the Letter of the Law," *Interior Landscape* 11(4): 35, 1994.

GRIFFITH, L.: "Tip Burn of *Dracaena Massangeana* Cane," *Florida Foliage* 11(6): 41, 43, 1985.

HAMLEN, R. A.: "Techniques of Pesticide Application and Regulation," *Proceedings of the 1977 National Tropical Foliage Short Course*, pp. 216–222.

HAMLEN, R. A., D. E. SHORT, and R. W. HENLEY: "Detect and Control Insects and Pests on Tropical Foliage Plants," *Florist* 9(3): 72–79, 1975.

HAMLEN, R. A., and M. V. WETTSTEIN: "An Evaluation of the Phytotoxicity of Enstar, Orthene and Pirimor on Twenty Species of Tropical Foliage Plants Grown under Greenhouse Conditions," *Foliage Digest* 1(4): 13–15, 1978.

HENLEY, R. W., ed.: *A Pictorial Atlas of Foliage Plant Problems*, Central Chapter, Florida Foliage Association, Apopka, Fla., 1983.

JOINER, J. N., ed.: *Foliage Plant Production*, Prentice-Hall, Inc., Englewood Cliffs, N.J., 1981.

KIPLINGER, D. C.: "Pointers on Diagnosing Problems with Foliage Plants," *Ohio Florists' Association Bulletin* 554: 5–8, 1975.

KNAUSS, J. F.: *Common Diseases of Tropical Foliage Plants: I. Foliar Fungal Diseases*, ARC—Apopka Research Report RH-75-6, IFAS, University of Florida, Apopka, Fla.

____: "Control of Basal Stem and Root Rot of Christmas Cactus Caused by *Phythium aphanidermatum* and *Phytophthora parasitica*," *Proceedings of the Florida State Horticultural Society* 88: 567–571, 1975.

____: "Description and Control of a Cutting Decay of Two Foliage Plant Species by *Rhizoctonia solani*," *Plant Disease Reporter* 57(30): 222–225, 1973.

____: "Description and Control of *Pythium* Root Rot on Two Foliage Plant Species," *Plant Disease Reporter* 56(3): 211–215, 1972.

____: "*Rhizoctonia* Blight of *Syngonium*," *Proceedings of the Florida State Horticultural Society* 86: 421–424, 1973.

____, C. A. CONOVER, and R. A. HAMLEN: "Measurement and Application Rates of Fungicides and Insecticides for Small Volume Pesticide Applications on Foliage Plants," *Florida Foliage Grower* 13(8): 8–9, 1976.

LINDQUIST, R. K.: "Insects and Mites on Interior Landscape Plants," *Ohio Florists' Association Bulletin* 634: 10–14, 1982.

____, and C. C. POWELL, JR.: "Mites on Indoor Plants," *Interior Landscape Industry* 2(9): 56–58, 1985.

____: "Selecting and Using Pesticides in Interiors," *Interior Landscape Industry* 1(3): 80–81, 1984.

MALMON, J. G.: "Biological Warfare," *Interior Landscape Industry* 2(6): 20–22, 25–32, 1985.

____: "Pesticide Propriety," *Interior Landscape Industry* 2(7): 42–50, 1985.

MERCHANT, M.: "Who You Gonna Call? Good Bugs!," *Interiorscape* 10(4): 26–27, 38–39, 1991.

OLKOWSKI, W., and H. OLKOWSKI: "Natural Control of Common Pests on Indoor Plants," *Horticulture* 54(10): 38–39, 1976.

____, and S. DAAR: "IPM for House Plants," *The IPM Practitioner* 2(10–11): 7–8, 1980.

OSBORNE, L. S.: "Don't Drench Plants with Pesticides Meant to Be Used as Foliar Sprays," *Interscape* 5(41): 19, 1983.

PAVONE, C. P.: "The Universal Symptom—Yellow Leaves," *House Plants and Porch Gardens* 3(2): 54–61, 1978.

"Pesticides on Clothes," *Interscape* 2(9): T4, 1980.

PIRONE, P. P.; *Diseases and Pests of Ornamental Plants*, 5th ed., John Wiley & Sons, Inc., New York, 1978.

POWELL, C. C.: "The Art and Science of Diagnosis," *Interscape* 4(24): 9–11, 1982.

____, and R. K. LINDQUIST: *Pest and Disease Control on Indoor Plants*, Bulletin 711, Cooperative Extension Service, The Ohio State University, Columbus, Ohio, 1983.

____: "Pesticides and Common Sense Chemistry," *Ohio Florists' Association Bulletin* 637: 8–9, 1982.

Professional Guide to Green Plants, Florists' Transworld Delivery Association, Southfield, Mich., 1976.

RICHARDSON, M. R.: "Get the Right Bang for Your Bug," *Grower Talks* 47(2): 62ff., 1983.

RICHMAN, S. M., and A. G. GENTILE: "Reported Phytotoxicity of Acaracides and Insecticides to Ornamentals," in *Floricultural Insects and Related Pests—Biology and Control Section I*, A. G. Gentile and D. T. Scanlon, eds., Cooperative Extension Service, University of Massachusetts, Waltham, Mass., pp. 45–52.

RIDINGS, W. H., and J. F. KNAUSS: "*Pythium* Root Rot of Devil's Ivy," *Florida Foliage Grower* 14(9): 1–2, 1977.

"Safe Storage and Handling of Pesticides," *Grounds Maintenance* 18(11): 50, 52, 1983.

Safe Use of Pesticides, U.S. Dept. of Agriculture, PA589, Washington, D.C., 1964.

SHALLA, T. A.: "New Causes of Plant Diseases Recognized," *Florida Foliage Grower* 14(9): 4–5, 1977.

Standards for Certification of Pesticide Applicators, Operations Division, Office of Pesticide Programs, U.S. Environmental Protection Agency, Washington, D.C.

STEINER, M. Y., and D. P. ELLIOTT: *Biological Pest Management for Interior Plantscapes*, 2nd Edition, Alberta Environmental Centre, Vegreville, Alberta, Canada, 1987.

STEWARD, V. B.: *Biological Control Agents for Interiorscapes*, Lecture, Longwood Gardens, May 27, 1993.

SWEET, K.: "Integrated Pest Management," *Ornamental Outlook* 4(3): 14–16, 18–19, 1995.

VITTAM, P.: "Biological Control: What's Going On?," *PPGA News* 24(3): 7, 10–11, 1993.

WANG, E. L. H., and I. W. HUGHES: "A Stem Gall of Cactus Induced by *Meloidogyne incognita*," *Plant Disease Reporter* 60(11): 906–907, 1976.

YEARY, R. A.: "Pesticide Application Safety," *Interior Landscape Industry* 1(3): 51–55, 1984.

$\sim 11 \sim$

Acclimatization

Acclimatization is adaptation of a species from one environment to another. It happens to all plants when they are moved from one location to another, such as from a greenhouse or field production area to the building interior, or even within a room. Since plants cannot move themselves from an unfavorable location, they must change to meet the conditions where they are. Depending upon the magnitude of the change, the process may be slow. Recognition of the need to acclimatize plants is not new. Cushing, in 1814, recommended that plants be acclimatized gradually to new conditions. The term is synonymous with hardening.

WHY ACCLIMATIZE?

Growers of foliage plants have provided an environment that maximizes growth—high light, high nutrition, plenty of water, and high temperatures and humidity. Plants moved directly from the favorable growing environment into the low-light, low relative-humidity conditions found indoors are subject to stresses which result in rapid loss of quality due to leaf drop, reduced growth, and even death of the plant.

Light intensity and water stress seem to have the most conspicuous effect in the acclimatization process. Leaves from sun-grown plants differ structurally and physiologically from shade leaves. They are smaller in size, thicker, have a more reflective surface, and are close together on the stem. Sun leaves have thicker cell walls, heavier cuticle, and may have multiple epidermal layers. The chloroplasts within the mesophyll are aligned vertically, as are the grana, a mechanism that reduces excessive heating and potential injury, but which reduces the efficiency of photosynthesis. Sun-grown plants are not efficient in photosynthesizing at typical light levels indoors and require higher light intensities.

Shade leaves are thinner and broader, with more chloroplasts aligned horizontally, to maximize light utilization. Grana, too, have a horizontal, open-stack arrangement in low light. Shade leaves have a higher nitrogen content and thinner cell walls, and usually a single epidermal layer. Plants grown in low light are not capable of high rates of photosynthesis. They are more photosynthetically efficient, however, making greater use of light, but at a slower synthetic rate.

When placed in low light, several complex changes occur in leaves formed in high light, partially transforming them to low-light leaves. The process takes 4 to 8 weeks or more, and during the acclimatization period, food reserves may be depleted and some leaves will drop (Figure 11.1). With time, there is an increase in the number of chloroplasts and the leaves become a darker green. The chloroplasts and the grana are reoriented and assume a position similar to shade leaves. The rate of respiration declines, thereby reducing carbohydrate requirements. Since the anatomy of leaves cannot change, a plant is not 100% acclimatized until all the leaves have been replaced.

Acclimatization during production will result in the highest-quality plants; acclimatized foliage plants have become the standard of the industry. Growers are producing light-acclimatized plants by two methods. One is to grow the plants under specific reduced-light intensities for their entire production period. Such plants are capable of transition to the indoor environment with little shock.

The second way is to grow the plants under high light, even full sun, and then convert them to low light during the late stages of production, prior to shipping. Large *Ficus* are usually grown outdoors exposed to full sun so that they develop sturdy trunks and branches quickly. Prior to shipping, they are placed in reduced light to produce shade-type leaves. Depending on the size of the tree, 4 months to over 1 year under shade may be required to minimize leaf drop. Sun leaves will drop indoors. Research in Florida has shown that the appearance of shade-grown *Ficus benjamina* and *Ficus retusa* 'Nitida' is better than converted trees, and they lose fewer leaves indoors. Plants also have lower LCP, lower carbohydrate in roots and leaves, more leaf chlorophyll, and longer indoor life.

Palms grown in full sun do not acclimatize quickly and may require 12 to 18 months of shade before being moved indoors. Palms can be grown from seed to finish under shade, a process that takes a long time and greatly increases plant cost. Sun-grown bamboo may require 6 months of shade before installation inside.

Shade-grown plants take longer to produce and are more expensive than either converted or nonacclimatized plants. Special shade houses are used to produce shade-grown and acclimatized (converted) plants in Florida and other areas (Figure 11.2). In each procedure, the light intensities maintained vary with the species and nutrition and irrigation are carefully controlled. One should purchase shade-grown or acclimatized plants if possible; otherwise, check with the grower as to the specific cultural procedures used and act accordingly.

Foliage plants are often shipped long distances from grower to consumer. Several days may elapse during which the plants are in total darkness with zero photosynthesis. Acclimatized plants withstand extended periods of dark better than do

(a)

(b)

Figure 11.1 (a) *Ficus benjamina* 10 weeks after transfer from full sun to artificial light at 25, 75, and 125 fc, 12 hrs a day, 7 days a week. Note deterioration of plant. (b) *Ficus benjamina*, all plants originally grown in sun, then moved to 80% shade for 8 weeks, then indoors for 10 weeks at 125, 75, 25 fc light 12 hrs a day, 7 days a week. Eight weeks of acclimatization at 80% shade enhanced the longevity of the plant indoors, as did light of greater intensity. [From C. A. Conover and R. T. Poole, "Acclimatization of Tropical Foliage Plants," George J. Ball, Inc., *Grower Talks* 39(6): 6–14, 1975.]

(a)

(b)

Figure 11.2 (a) Structures for acclimatizing plants to reduced-light environments; (b) Large *Ficus* and palms undergoing acclimatization.

nonacclimatized plants. After 6 days in dark storage, there was no loss in quality of *Brassaia actinophylla, Philodendron scandens* 'oxycardium', and *Aphelandra squarrosa* 'Dania' 1 month after placement in a greenhouse at 10 kilolux (100 fc) light maximum. *Ficus benjamina* grown at low light intensity with low fertilizer levels dropped fewer leaves than did plants grown at high light and fertilizer levels. To minimize loss of quality that may result from extended shipping, all plants should be unwrapped, watered if necessary, and placed in light of the proper intensity immediately upon receipt.

Changes in relative humidity create large changes in water demand. In building interiors, the humidity is usually low and favors rapid transpiration. If the roots are not extensive enough to supply sufficient water to satisfy the needs of transpiration, water stress occurs, stomata close, photosynthesis stops, and the plant may wilt. The plant draws on reserve food, and deterioration begins. If the area or efficiency of the root system is reduced, as in transplanting or repotting, absorptive surface is lost and a water deficit develops almost immediately. There is also a demand for food for initiation and growth of new roots. In response, old leaves senesce, reducing water loss, and making metabolites mobilized from old leaves available for new growth. Eventually, the systems will come into balance. Thus, the effects of water stress must be minimized by the acclimatization process.

There is also evidence of a need for acclimatization of the growing medium, and consideration should be given to soluble-salt levels at the time plants are moved indoors. Working with *Aphelandra*, Conover and Poole (1977) found that the best plants after 8 weeks indoors under 0.8 kilolux of cool-white fluorescent light for 12 hrs daily were those produced on the lowest fertilizer level, or those watered daily at any fertilizer level. *Ficus benjamina* produced at the highest fertilizer levels had the greatest leaf drop after 6 months indoors.

If repotting is necessary, it should be done as soon as the plants are received, because acclimatization to the new growing medium is also needed.

FACILITIES NEEDED

Acclimatization will be required of all plants not totally acclimatized by the grower. It makes no difference whether the plants were produced in Florida, Texas, California, or in a northern greenhouse—they must be acclimatized before they go into the retail shop or interior planting. A special holding facility should be available. A shaded greenhouse, lighted building, or room equipped for environment control is satisfactory for the acclimatization process (Figure 11.3).

PROCEDURE

The first step should be to learn from the grower as much as you can about the cultural practices. What light levels were used? Fertilizer program? Pesticides used?

Figure 11.3   Holding/acclimatization facility.

Remove any Osmocote, Mag-Amp, or other slow-release fertilizer which may be on the surface of the medium and thoroughly leach. High levels of nutrition have been used in production, but indoors, they will reduce survival rates and retard acclimatization of plants. High salt levels cause loss of plant quality due to chlorosis (yellowing) and necrosis (death) of leaves, and leaf drop. The entire plant may die.

Additional fertilizer should not be applied for 2 to 3 months unless a soil test indicates otherwise. Fertilization in the retail shop should not be necessary.

Following leaching, watering frequency should be reduced and the medium should be kept moderately moist.

The light intensity must gradually be reduced from the 6,000 to 10,000 fc of the production area to less than 150 fc found in most building interiors. If there has been no previous acclimatization, place the plants in 50% shade (3,000 to 5,000 fc) for several weeks, then further reduce the light intensity for an additional period of time. If using natural light, reduction of light intensity to 1,500 to 2,000 fc should be adequate. In a holding room, 300 to 500 fc of artificial light from either fluorescent or HID lamps, 12 hrs daily, 7 days a week is the minimum recommendation. Following 6 to 8 weeks of treatment, the plants may be moved into the building interior, where sun-grown plants should receive a minimum of 150 fc, 12 hrs daily, 7 days a week, and shade-grown, 75 to 100 fc. In a plant shop, 150 to 200 fc is recommended.

If some acclimatization has been provided, place the plants under 300 to 500 fc cool-white fluorescent for 12 hrs daily, 7 days a week until fully acclimatized in 2 to 3 weeks. Plants may then go into the indoor garden under the minimum light intensities and durations listed above.

In conjunction with light-intensity acclimatization, the relative humidity should be reduced gradually from the 85 to 95% found in the production area to the 25 to 40% of building interiors. Temperature should also be lowered gradually to that indoors.

At the outset of the acclimatizing period, all plants should be inspected for insects and other pests and diseases. Any problems should be corrected before the plants leave the area. Necessary pruning and shaping should be done during acclimatization, and the foliage cleaned.

TIME REQUIRED

The length of time required for a plant to become acclimatized varies with the species, its previous treatment, and the magnitude of the change. From 1 to 8 or 10 weeks or more may be necessary. *Philodendron* and *Aglaonema* seem to require little, if any. Sun-grown *Ficus benjamina* and *Brassaia* acclimatize in 40 to 80% shade in about 5 weeks, and *Ficus nitida* requires 10 weeks at 50% shade. The larger the plant, the longer the process takes. To acclimatize properly, a plant must have a fully developed root system.

If the grower, wholesaler, retailer, or landscaper does not acclimatize the plants, it will occur with the customer, and undesirable side effects may occur. Leaf drop is the primary symptom. In low light, plants will drop leaves until the rate of food use is at least equal to its manufacture. This is the plant's light compensation point. In extremely low light, all the leaves may drop over a period of several weeks, and the plant will die. Since food is not available, new growth cannot occur. The quality of nonacclimatized plants declines rapidly indoors and their aesthetic value is greatly reduced. Clients become dissatisfied and replacement of plants is necessary at an early date.

SUMMARY

Acclimatization is adaptation of a species to a new environment. It occurs when plants are moved from the production environment to indoors, or within a room. Environmental changes cause stresses that may result in leaf drop, reduced growth, and even death of the plant.

Adaptation to low light intensity and low humidity are most important. Unless shade-grown or acclimatized plants are used, the plants should be subjected to gradually reduced light intensities over a period of several weeks to months before they are sold or used indoors. Thoroughly leach the growing medium, water less frequently, and gradually reduce the relative humidity. Pests and diseases should be eradicated during acclimatization, and the plants pruned, shaped, and cleaned. Several weeks may be required, depending upon the plant species.

REFERENCES

BIAMONTE, R. L.: "Nutrition and Light in Plant Acclimatization," *Florists' Review* 163(4223): 118–121, 1978.

BJORKMAN, O., and P. HOLMGREN: "Adaptability of the Photosynthetic Apparatus to Light Intensity in Ecotypes from Exposed and Shaded Habitats," *Physiologia Plantarum* 16: 889–914, 1963.

BOODLEY, J. W.: "Acclimatization," Lecture, 1978 National Tropical Foliage Short Course, Orlando, Fla.

____: "Acclimatization: The Critical First Step for Indoor Gardening Success," *Horticulture* 54(10): 34–35, 1976.

____: "Foliage Plant Acclimatization," *Ohio Florists' Association* 562: 5, 1976.

BROSCHAT, T. K., and H. DONSELMAN: "Light Acclimatization in *Ptychosperma elegans*," *HortScience* 24(2): 267–268, 1989.

CONOVER, C. A. "Plant Conditioning," *American Nurseryman* 148(4): 16, 130–133, 1978.

____: "Plant Conditioning—It's Here," *New Horizons from the Horticultural Research Institute 1978*, Horticultural Research Institute, Inc., 230 Southern Building, Washington, D.C. 20005.

____, and R. T. POOLE: "Acclimatization of *Aphelandra squarrosa* Nees cv Dania," *HortScience* 12(6): 569–570, 1977.

____: "Acclimatization Revisited," *Foliage Digest* 13(5): 6–7, 1990.

____: "Acclimatizing Your Tropical Foliage Plants," *Floral and Nursery Times* 3(3): 2ff., 1981.

____: "Effects of Cultural Practices on Acclimatization of *Ficus benjamina* L.," *Journal of the American Society for Horticultural Science* 102(5): 529–531, 1977.

____: "*Ficus benjamina* Leaf Drop," *Florists' Review* 151(3925): 29, 67, 1973.

____: "Light and Fertilizer Recommendations for Production of Acclimatized Potted Foliage Plants," *Foliage Digest* 13(6): 1–6, 1990.

CUSHING, J.: *The Exotic Gardener*, W. Bulmer and Co., London, 1814.

JOHNSON, C. R., D. L. INGRAM, and J. E. BARRETT: "Effects of Irrigation Frequency on Growth, Transpiration, and Aclimatization of *Ficus benjamina* L., *HortScience* 16(1): 80–81, 1981.

JOHNSON, C. R., J. K. KRANTZ, J. N. JOINER, and . A. CONOVER: "Light Compensation Point and Leaf Distribution of *Ficus benjamina* as Affected by Light Intensity and Nitrogen-Potassium Nutrition," *Journal of the American Society for Horticultural Science* 104(3): 335–338, 1979.

MCWILLIAMS, E. L., and A. K. MCWILLIAMS: "Leaf Anatomy and Ultrastructure of Acclimatized and Non-acclimatized *Peperomia obtusifolia*" (abstract), *HortScience* 13(3): 344, 1978.

PASS, R. G., and D. E. HARTLEY: "Net Photosynthesis of Three Foliage Plants under Low Irradiation Levels," *Journal of the American Society for Horticultural Science* 104(6): 745–748, 1979.

PETERSON, J. C.: "The Impact of Leaf Anatomy on *Ficus* Survival Indoors," *Interior Landscape Industry* 3(4): 30–35, 1986.

POOLE, R. T., and C. A. CONOVER: "Influence of Shade and Nutrition during Production and Dark Storage Simulating Shipment on Subsequent Quality and Chlorophyll Content of Foliage Plants," *HortScience* 14(5): 617–619, 1979.

RICHARDSON, D.: "Acclimatization and Job Success," *Interiorscape* 6(3): 26, 28–33, 1984.

VLAHOS, J., and J. W. BOODLEY: "Acclimatization of *Brassaia actiophylla* and *Ficus nitida* to Interior Environmental Conditions," *Florists' Review* 154(3989): 18–19, 56–60, 1974.

WETHERILL, D. F.: "Condition Plants for Interior Use," *Flower News* 32(1): 41ff., 1978.

~12~

Recommended Plants

There are more than 300 species and cultivars of foliage plants grown commercially. Many excellent publications exist which give cultural information for specific foliage and other indoor plants, and no attempt will be made to duplicate that material here.

Some foliage plants are more adaptable than others to the indoor environment, and Table 12.1 lists recommended plants for interior plantscaping. The list is not all-inclusive, but includes species that have performed well for extended periods indoors (Figures 12.1 to 12.51).

The recommended minimum light intensities are for at least 12 hrs every day, and represent the best information available at this time. Considerable variation exists in the literature regarding light requirements for interior plants because little research has been done to date to determine the light-compensation points for each foliage plant species and cultivar. Many of the plants listed will tolerate intensities as low as 20 to 25 fc, but the quality of the plants in low light will decline faster than similar plants kept at, or above, the minimum level, and will require frequent replacement. For all plants, brighter light, within limits, is beneficial. With brighter light, the hours of the lighting period may be reduced, and the plants will remain attractive for a longer period of time.

Tropical and subtropical plants are most frequently used indoors. The planting at the Ford Foundation in New York City (Figure 12.52), installed in 1967, is probably the first example of an interiorscape using temperate species. A major project for its time, and still a landmark in the city, the garden originally contained such plants as *Magnolia grandiflora*, *Jacaranda*, *Cryptomeria japonica*, *Pachysandra*, and *Indica azaleas*. Some survived, others did not, and the temperate plants were gradually phased out. A recent renovation has seen elimination of the last of the *Magnolia*; majestic *Podocarpus* have taken their place. The original design intents of

TABLE 12.1 Recommended Plants for Interior Plantscaping

Plant Name (and Figure No.)	Common Name	Family	Origin	Uses[a]	Minimum Light Requirement fc	Minimum Light Requirement Light Level[b]	Tolerate 50°F Temperature	Moisture Dry	Moisture Moist
Aechmea fasciata (12.1)	Silver vase plant	Bromeliaceae	South America	2, 3	100	M		X	
Agave americana (12.2)	Century plant	Agavaceae	Mexico	5	150	H		X	
Aglaonema (12.3) (all cultivars)	Chinese evergreens	Araceae	Asia	2, 3, 4	50[c]	L		X	
Anthurium hookeri (12.4)	Bird's-nest anthurium	Araceae	West Indies	4	50	L			X
Anthurium sp. (12.5)	Flamingo flower	Araceae	Colombia, Peru	1, 2, 3, 4	75	M			X
Araucaria bidwilli (12.6)	Monkey-puzzle tree	Araucariaceae	Australia, Queensland	2, 4, 5	100	M	X		X
A. heterophylla (1.4)	Norfolk Island pine	Araucariaceae	Norfolk Island	2, 4, 5	100	M	X		X
Asparagus densiflorus 'Sprengeri' (6.11)	Asparagus fern	Liliaceae	South Africa	1, 2, 3, 8	100	M	X	X	
Aspidistra elatior (12.7)	Cast iron plant	Liliaceae	China	3, 4	50[c]	L	X		
Beaucarnea recurvata (12.8)	Pony-tail palm	Agavaceae	Mexico	1, 4, 5	100	M	X		X
Brassaia actinophylla (1.5, 6.9)	Schefflera, umbrella tree	Araliaceae	Queensland, Java, New Guinea	2, 4, 5	100[c]	M		X	
Bucida buceras	Black olive	Combretaceae	Florida, West Indies to Panama	4, 5	300	H			X
Calathea sp. (12.9)	Calathea	Marantaceae	Brazil, Venezuela, Peru	1, 2, 3, 4	75	M			X
Caryota mitis (12.10)	Dwarf fishtail palm	Palmae	Burma, Indonesia	4, 5	75	M			X
Cereus peruvianus (12.11)	Column cactus	Cactaceae	Possibly hybrid	1, 4	100	M		X	
Chamaedorea elegans (1.7, 12.12)	Parlor palm	Palmae	Mexico	1, 2, 4	50[c]	L			X

(continued)

TABLE 12.1 (Continued)

Plant Name (and Figure No.)	Common Name	Family	Origin	Uses[a]	Minimum Light Requirement fc	Light Level[b]	Tolerate 50°F Temperature	Moisture Dry	Moist
C. erumpens (6.5, 6.8)	Bamboo palm	Palmae	Honduras	2, 4, 5	50	L			X
C. seifrizii	Reed palm	Palmae	Mexico, Yucatan	4, 5	50	L			X
Chamaerops humilis (12.13)	Fan palm	Palmae	Mediterranean, Spain, Morocco	4, 5	100	M	X		X
Chlorophytum comosum 'vittatum' (6.12)	Spider plant	Liliaceae	South Africa	1, 8	100	M	X		X
Chrysalidocarpus lutescens (12.14)	Areca palm	Palmae	Madagascar (Malagasy Republic)	4, 5	100	M			X
Cissus antarctica (12.15)	Kangaroo vine	Vitaceae	Australia, New South Wales	1, 8	75	M	X		X
C. rhombifolia (1.8)	Grape ivy	Vitaceae	West Indies, Northern South America	1, 8	75	M			X
C. rhombifolia 'Ellen Danica' (12.16)	Oak-leaf ivy	Vitaceae	Mutant	1, 8	75	M			X
Codiaeum variegatum (cultivars) (6.9)	Croton	Euphorbiaceae	Southern India to Indonesia	1, 2, 4, 5	100	M-H			X
Coffea arabica	Coffee	Rubiaceae	Ethiopia, Mozambique, Angola	2, 4, 5	100	M			X
Cordyline terminalis 'Baby Doll' (12.17)	Ti plant	Agavaceae	India to Polynesia	2, 4, 5	100	M			X
Crassula argentea (12.18)	Jade plant	Crassulaceae	South Africa, Natal	1, 4	50[c]	M	X		X
Cycas circinalis	Fern palm	Cycadaceae	Madagascar, southern India to the Philippines	5	100	M			X

Name	Common name	Family	Origin						
C. revoluta (12.19)	Sago palm	Cycadaceae	Southern Japan to Java	5	100	M			X
Dieffenbachia (1.5, 12.20) (all cultivars)	Dumbcane	Araceae	Tropical America	1, 2, 3, 4, 5	100[c]	M		X	
Dizygotheca elegantissima (12.21)	False aralia	Araliaceae	New Hebrides	2, 4, 5	100	M		X	
Dracaena deremensis 'Janet Craig' (1.5, 1.7, 6.7)	Janet Craig Dracaena	Agavaceae	Mutant	1, 2, 3, 4, 5	75	M		X	
D. deremensis 'Warneckii' (1.5, 1.7, 6.9)	Striped dracaena	Agavaceae	Tropical Africa	1, 2, 3, 4, 5	75[c]	M		X	
D. fragrans	Green corn plant	Agavaceae	Tropical Africa	2, 4, 5	75	M		X	
D. fragrans 'massangeana' (6.10)	Corn plant	Agavaceae	Tropical Africa	2, 4, 5	75	M		X	
D. marginata (1.1, 6.8)	Madagascar dragon tree	Agavaceae	Madagascar	2, 4, 5	75[c]	M		X	
D. reflexa (12.22)	Malaysian dracaena	Agavaceae	Madagascar, India	1, 4	75	M		X	
D. sanderana (12.23)	Ribbon plant	Agavaceae	Cameroons, Congo	1, 2, 4	75	M		X	
Epipremnum aureum (6.12, 12.24)	Pothos	Araceae	Solomon Islands	1, 6, 7, 8	75[c]	M			X
Euphorbia lactea	Candelabra plant	Euphorbiaceae	India, Sri Lanka	4	200	H			X
Fatsia japonica (6.5)	Japanese aralia	Araliaceae	Japan	1, 4, 5	75	M			X
Ficus benjamina (1.5, 12.57, 12.58)	Weeping fig	Moraceae	India, Malaya	4, 5	75	M	X	X	
F. elastica 'Decora' (1.1, 6.11)	Rubber plant	Moraceae	India, Malaya	2, 4, 5	75[c]	M		X	
F. elastica 'Doescheri' (12.25)	Variegated rubber plant	Moraceae	Sport of F. elastica	2, 4, 5	75	M		X	
F. lyrata (12.26)	Fiddle-leaf fig	Moraceae	Tropical West Africa	2, 5	75	M		X	

(continued)

TABLE 12.1 (Continued)

Plant Name (and Figure No.)	Common Name	Family	Origin	Uses[a]	Minimum Light Requirement		Tolerate 50°F Temperature	Moisture	
					fc	Light Level[b]		Dry	Moist
F. macellandi 'Alii' (12.27)	Alii	Moraceae		4, 5	75	M	X		X
F. repens (pumila)	Creeping fig	Moraceae	China, Japan, Australia	3, 8	50	L			X
F. retusa 'Nitida' (6.3, 6.7)	Indian laurel	Moraceae	Malaya	4, 5	75	M			X
Hedera helix (12.28)	English ivy	Araliaceae	Europe, Asia, North Africa	1, 8	50	L	X		X
Heliconia sp.	Lobster claws	Heliconiaceae	Tropical America	3	200	H			X
Homalomena wallisii (12.29)	Silver shield	Araceae	Colombia	1, 3, 4	30	L			X
Howea forsterana (6.7)	Kentia palm	Palmae	Lord Howe Islands	4, 5	50	L	X		X
Hoya carnosa (12.30)	Wax plant	Asclepiadaceae	Queensland, southern China	1, 6	100	M		X	
Livistona rotundifolia	Chinese fan palm	Palmae	Southern China	4, 5	100	M			X
Neodypsis decaryi	Triangle palm	Palmae	Madagascar	4, 5	100	M			X
Neoregelia carolinae 'tricolor' (12.31)	Striped blushing bromeliad	Bromeliaceae	Brazil	1	100	M			X
Nephrolepis exaltata 'Bostoniensis' (12.32)	Boston fern	Polypodiaceae	Mutant	1, 3, 4	75	M	X		X
N. 'Dallasi'	Dallas fern	Polypodiaceae	Mutant	1, 3, 4, 8	75	M	X		X
N. 'Florida Ruffles' (6.12)	Dwarf feather	Polypodiaceae	Mutant	1, 3, 4	75	M	X		X
Philodendron bipenni-folium (12.33)	Fiddle-leaf philodendron	Araceae	Southern Brazil	6, 7	75	M			X

Botanical name	Common name	Family	Origin		Max ht				
P. 'Emerald Queen' (2.7, 6.5)	Emerald queen philodendron	Araceae	Hybrid	3, 6, 7	75	M			X
P. pertusum (1.1) (*Monstera deliciosa*)	Split-leaf philodendron	Araceae	Southern Mexico, Guatemala	2, 6, 7	75	M			X
P. 'Pluto' (12.34)	Pluto	Araceae	Hybrid	4, 5	75	M			X
P. scandens 'oxycardium' (12.35)	Heart-leaf philodendron	Araceae	Puerto Rico to Jamaica and Central America	1, 6, 7	75c	M			X
P. selloum (12.36)	Lacy-tree philodendron	Araceae	Southwestern Brazil	5	75c	M			X
P. 'Xanadu' (12.37)	Xanadu	Araceae	Hybrid	3, 4, 5	75	M			X
Phoenix roebelenii (12.38)	Dwarf date palm	Palmae	Assam, Burma, Vietnam	1, 4, 5	50c	L			X
Phyllostachys nigra (1.6, 12.39)	Black bamboo	Gramineae	South China	5	500	H	X		
Pittosporum tobira (12.40)	Japanese Pittosporum	Pittosporaceae	China, Japan	4, 5	75c	M	X	X	
Pleoblastus viridistriatus	Dwarf green bamboo	Gramineae	Japan	3, 4	500	H	X		X
Podocarpus gracilior	Weeping podocarpus	Podocarpaceae	Kenya, Uganda, Ethiopia	4, 5	100	M	X		X
Podocarpus macrophyllus (12.41)	Japanese yew	Podocarpaceae	China, Japan	1, 4, 5	100	M	X		X
Polyscias balfouriana	Dinner plate aralia	Araliaceae	New Caledonia	1, 4, 5	75	M			X
P. fruticosa (12.42)	Ming aralia	Araliaceae	India, Malaysia, Polynesia	1, 4, 5	75c	M			X
Ptychosperma elegans (1.5)	Alexander palm	Palmae	Queensland	4, 5	200	H			X
Ravenala madagascariensis	Traveler's tree	Musaceae	Madagascar	4, 5	100	M			X
Ravenea rivularis (12.43)	Majesty palm	Palmae	Madagascar	4, 5	300	H	X		X
Rhapis excelsa (12.44)	Lady palm	Palmae	Southern China	4, 5	50	M	X		X
Sansevieria trifasciata 'laurentii' (12.45)	Variegated snake plant	Agavaceae	Northeast Congo	1, 4	50	L		X	
Schefflera arboricola (12.46)	Dwarf schefflera	Araliaceae	India, Australia, Philippines	2, 4	50	L			X

(continued)

TABLE 12.1 (Continued)

Plant Name (and Figure No.)	Common Name	Family	Origin	Uses[a]	Minimum Light Requirement fc	Minimum Light Requirement Light Level[b]	Tolerate 50°F Temperature	Moisture Dry	Moisture Moist
Spathiphyllum (all cultivars) (12.47)	White flag	Araceae	Tropical America	3, 4	75[c]	M			X
Strelitzia nicolai (12.48)	White bird-of-paradise	Musaceae	Africa	4, 5	100	M			X
Syngonium podophyllum (1.3, 12.49)	Nephthytis	Araceae	Mexico to Costa Rica	1, 6, 7	75	M			X
Syagrus romanzoffianum	Queen palm	Palmae	So. Brazil to W. Argentina, Bolivia	4, 5	200	H			X
Veitchia merrilii (12.50)	Christmas palm	Palmae	Philippines	1, 4, 5	100[c]	M			X
Washingtonia robusta (1.9)	Mexican fan palm	Palmae	N.W. Mexico, Sonora, Baja Calif.	4, 5	200	H			X
Yucca elephantipes (6.6)	Joshua tree	Agavaceae	Southeastern Mexico, Guatemala	1, 4, 5	100	M	X	X	
Zamia sp. (12.51)	Cardboard palm	Cycadaceae	Florida, Mexico, Colombia	2, 4, 5	100	M			X

[a]Key to uses: 1, small pot specimen for table or desk; 2, accent plant for planter; 3, filler for planter; 4, small specimen; 5, large specimen; 6, small totem pole (18 to 24 in.); 7, large totem pole (36 to 48 in.); 8, hanging basket.

[b]L, 3 W/m² irradiance; M, 9 W/m²; H, 24 W/m².

[c]Will tolerate light intensities of 20 to 25 fc.

Figure 12.1 *Aechmea fasciata*, silver vase plant.

Figure 12.2 *Agave americana*, century plant.

the garden have been restored, with tropical and subtropical plants creating the temperate look of the original garden.

To succeed indoors, temperate species must be able to adapt to low light and maintain vigor without experiencing a natural dormancy period when plants shed their leaves. By maintaining long day conditions and constant warm temperatures, both common in shopping malls and other indoor spaces, dormancy can be prevented or delayed. Many dormant plants need an extended cold period to break dormancy. Of course, satisfying this requirement is impossible in the interior landscape, as people's comfort needs must be met first. The response is species-dependent.

Studies by Keever and Cobb show that temperate species such as *Podocarpus, Gardenia, Pittosporum, Mahonia,* and *Fatsia* can be successfully used indoors. Friendship, Tsujita, and Ormrod report that selected species of *Caragana* (Pea-Tree), *Cornus, Rhus, Ilex, Taxus,* and *Euonymus* have promise for interior landscaping.

(a)

(b)

(c)

(d)

Figure 12.3 (a) *Aglaonema modestum*, Chinese evergreen; (b) *A. commutatum* 'White Rajah', golden evergreen; (c) *A. commutatum* 'Silver Queen'; (d) *A. commutatum* 'Emerald Beauty' ('Maria').

Figure 12.6 *Araucaria bidwilli*, monkey-puzzle tree.

Figure 12.4 *Anthurium hookeri*, bird's-nest anthurium.

Figure 12.7 *Aspidistra elatior*, cast iron plant.

Figure 12.5 *Anthurium* sp. 'Lady Jane', Lady Jane anthurium.

Figure 12.8 *Beaucarnea recurvata*, pony-tail palm.

Figure 12.9 *Calathea roseopicta*.

Figure 12.10 *Caryota mitis*, fishtail palm.

Figure 12.11 *Cereus peruvianus*, column cactus.

Figure 12.12 *Chamaedorea elegans*, parlor palm.

Figure 12.13 *Chamaerops humilis*, fan palm.

Figure 12.14 *Chrysalidocarpus lutescens*, areca palm.

Figure 12.15 *Cissus antarctica*, kangaroo vine.

Figure 12.16   *Cissus rhombifolia* 'Ellen Danica', oak-leaf ivy.

Figure 12.17   *Cordyline terminalis* 'Baby Doll', ti plant.

Figure 12.18   *Crassula argentea*, jade plant.

Figure 12.19 *Cycas revoluta*, sago palm.

(b)

(a)

(c)

Figure 12.20 (a) *Dieffenbachia maculata* 'Camille'; (b) *D. amoena* 'Alix'; (c) *D. maculata* 'Exotica compacta', dumbcane.

Figure 12.21 *Dizygotheca elegantissima,* false aralia.

Figure 12.22 *Dracaena reflexa,* Malaysian dracaena.

Figure 12.23 *Dracaena sanderana,* ribbon plant.

Figure 12.24 *Epipremnum aureum* 'Marble Queen', Marble Queen pothos.

Figure 12.25 *Ficus elastica* 'Doescheri', variegated rubber plant.

Figure 12.26 *Ficus lyrata*, fiddle-leaf fig.

Figure 12.27   *Ficus maclellandi*, 'Alii'.

Figure 12.28   *Hedera helix*, English ivy.

Figure 12.29   *Homalomena wallisii*, silver
shield.

Figure 12.30   *Hoya carnosa varietata;*
variegated wax plant.

Figure 12.31 *Neoregelia carolinae* 'tricolor', striped blushing bromeliad.

Figure 12.32 *Nephrolepis exaltata* 'Bostoniensis', Boston fern.

Figure 12.33 *Philodendron bipennifolium*, fiddle-leaf philodendron.

Figure 12.34   *Philodendron* 'Pluto'.

Figure 12.35   *Philodendron scandens* 'oxycardium', heart-leaf philodendron.

Figure 12.36   *Philodendron selloum*, lacy-tree philodendron.

Figure 12.37   *Philodendron* 'Xanadu'.

Figure 12.38   *Phoenix roebelenii*, dwarf date palm.

Figure 12.39   *Phyllostachys nigra*, black bamboo.

Figure 12.40   *Pittosporum tobira*,
Japanese pittosporum (variegated form).

Figure 12.41   *Podocarpus macrophyllus*,
Japanese yew.

Figure 12.42   *Polyscias fruticosa*, Ming
aralia.

Figure 12.44 *Rhapis excelsa*, lady palm.

Figure 12.43 *Ravenea rivularis*, majesty palm.

Figure 12.45 *Sansevieria trifasciata* 'laurentii', variegated snake plant.

Figure 12.46 *Schefflera arboricola,* dwarf schefflera.

Figure 12.47 *Spathiphyllum* sp., white flag.

Figure 12.48 *Strelitzia nicolai,*
white bird-of-paradise.

Figure 12.49 *Syngonium
podophyllum* 'White Butterfly',
nephthytis.

Figure 12.50 *Veitchia merrilii,* Christmas palm.

Figure 12.51 *Zamia* sp., cardboard palm.

Figure 12.52 Section of the original planting at the Ford Foundation, New York.

Certain species of *Acer*, *A. platanoides*, and *A. palmatum* have short dormant periods and have potential indoors.

Keever, Cobb, and Stephenson demonstrated that *Euonymus* sp., *Hosta*, *Liriope*, *Magnolia*, *Mahonia*, *Nandina*, *Ophiopogon* (Lily-Turf), *Ternstroemia*, *Aucuba*, *Ligustrum*, *Illicium*, and *Raphiolepis* (Indian Hawthorn) all performed well indoors for 15 weeks.

When placed in a large planting, dormant plants are less conspicuous. If they have an attractive form and interesting bark patterns and color when leafless, dormant plants could enhance rather than detract from the landscape. It is suggested that before making a commitment to a large-scale planting using temperate species, a 12–18 month trial using a few plants be made.

In an effort to simulate a Minnesota woodland and broaden the list of plants suitable for use indoors, temperate species have been installed in Knotts' Camp Snoopy at Mall of America on an experimental basis. A 1993 survey reported by Clein showed that 14 species gave good to excellent results after 1 year. The successful plants were *Acer palmatum*, *Acer rubrum*, *Betula nigra*, *Camellia* sp., *Cercis canadensis*, *Cornus florida*, *Corynocarpus laevigata*, *Grevillea robusta*, *Lagerstroemia indica*, *Ligustrum japonicum*, *Pinus canariensis*, *Pinus thunbergii*, *Prunus caroliniana*, and *Sequoia sempervirens*. Evaluations are continuing and will be reported in the future.

FLOWERING PLANTS

It is commonplace for interiorscapers to use flowering potted plants throughout the year, or seasonally. Color is the main reason. As a mass of color or an accent to foliage displays, flowering plants add interest and excitement.

People respond emotionally to blooming potted plants, and color can be used to foster specific responses. For example, red is a bold, aggressive, attention-getting color, while blue is a restful hue. Yellow is a cheerful color eliciting happy, warm thoughts. Because blooming potted plants produce emotional responses which range from comfortable to productive to cheerful, and even to aggressive, colors must be chosen carefully.

Use of flowering plants must be planned in advance. Not all work well in all locations. Selection depends on the site conditions, plus cost, availability, seasonality, maintenance requirements, and other factors. Because the blossoms are short-lived, flowering plants are not grown in place. They are forced in greenhouses and installed in bloom. Plants should be ordered well in advance, as special growing may be required. One should purchase high-quality, acclimatized plants and select cultivars appropriate to the design scheme and suited to the intended use. To add days to the display, install the plants at the proper stage of flower development.

Flowering plants are used on a regular basis wherein they are changed periodically (rotational planting) and for seasonal display (Figure 12.53). To facilitate installation of blooming plants in a rotational planting, double-pot. In double-potting, a plant in its grow pot is placed into another, slightly larger pot that has been

Figure 12.53   Poinsettia tree.

permanently positioned in the planter bed or decorative container. The ready-made opening makes replacement quick and easy. The old plant is removed and the new one slipped in its place without disturbing the roots of the permanent planting. Where subirrigation planters are used, install the plants as suggested by the manufacturer. To do otherwise will cause the system to malfunction and shorten the plant's life. To minimize the mess associated with replacement, and clean-up time, bring plastic bags and drop cloths.

When plants arrive from the grower, open the cartons and unsleeve the plants immediately. This eliminates potential injury from ethylene and will extend display life. Once installed, flowering plants require a different approach to maintenance than foliage plants. Maintenance is more demanding, hence, more expensive. Flowering plants must be replaced frequently. They need more water than foliage plants, so visits to the site may have to be made two or three times a week to replenish the supply. Use of subirrigation planters will help eliminate the extra trips. Mulching also reduces frequency of watering. Do not allow the plants to wilt. Application of fertilizer is unnecessary.

Adequate light is essential. Bright, indirect light of 200 fc or more is recommended. Low light causes rapid decline. Space the plants far enough apart so that the lower leaves receive adequate light. Premature chlorosis is thus prevented. Lamps may distort the flower colors. Cool-white fluorescent, for example, while fine for whites and yellows, causes bronze, red, pink, and lavender to appear faded. Deluxe white fluorescent or incandescent spot lamps will enhance the appearance of these colors.

Flower life will be maximized at temperatures of 55 to 70°F (13 to 21°C). If the environment is too warm, early senescence will result. The leaves dry and fall,

Figure 12.54 Faded flowers detract from a planting. Note the grow pots, which should have been hidden by the mulch.

flowers may not open completely or abort, and buds may blast. A reduction of temperature at night would be desirable but is not possible in many building interiors. Humidity also affects the longevity of the blooms. Bud drop and failure of buds to open may be indicative of low humidity.

Flowering plants must always look their best. To keep them attractive, groom them on a regular basis. Remove faded or spent blooms, yellowing leaves, and broken stems. Unsightly or dead plants should be removed or replaced immediately (Figure 12.54).

Depending on the species chosen, potted flowering plants will last a few days to 3 or 4 weeks in the plantscape. Generally speaking, the plants themselves are inexpensive; however, maintaining flowering plants is labor intensive and therefore costly. The client must understand this additional investment and budget accordingly. He or she must also recognize the limitations of various flowering plants in terms of size, color, and longevity.

Not all flowering plants or cultivars of the same species are satisfactory indoors. Learn and use those which work best for you.

Potted chrysanthemums are excellent indoors and form the basis for most programs (Figure 12.55). They are available year-round in a wide color range and a variety of flower forms. Properly maintained, mums remain attractive for 3 to 4 weeks. To maximize chrysanthemum longevity, select long-lasting cultivars, buy plants grown to last, have plants delivered at the proper stage of development, and practice proper horticultural techniques on site. Flower longevity varies by cultivar.

Figure 12.55 Potted chrysanthemums are excellent indoors and form the basis for most programs using flowering plants.

Learn which cultivars last longest and order them from your supplier by name. Some long-lasting chrysanthemum cultivars are listed in Table 12.2.

How the plants were grown in the greenhouse and the environment in shipping influence plant quality and longevity. Consult with your grower about the techniques used, especially light intensity and fertilizer program. Pot mums should be received and installed when the flowers are half opened. More or less than the 50% open stage decreases longevity.

Poinsettias are important plants for Christmas displays. Although some cultivars retain their leaves well, even after 30 days indoors, bract color usually fades due to relatively warm indoor temperatures. Consideration should be given to replacing the plants after 3 weeks. Although an added expense for the client, poinsettias installed just before Thanksgiving will begin looking bad by the middle of December. When ordering poinsettias for commercial plantings, cultivars that do not "go to sleep" at night (exhibit petiole droop) should be chosen.

Bromeliads have become important in the interiorscape. As a group, they are among the most versatile, longest-lasting, easiest-to-care-for choices for providing color. A wide variety of bromeliads is available, each having its own distinctive combination of flower and leaf colors and texture. Varieties of *Aechmea* (Figure 12.1), *Guzmania, Neoregelia* (Figure 12.31), *Tillandsia,* and *Vriesea* are recommended. Bromeliads require little maintenance; they do not wilt or drop leaves and

TABLE 12.2   Long-Lasting Potted Mums for Interior Landscapes

White	Yellow	Pink	Bronze
Boaldi	Bright Golden Anne	Blush	Cherry Pomona
Claro	Carnival	Capri	Cirbronze
Dana	Cartago	Charm Pomona	Dark Bronze Charm
Dare	Cream Boaldi	Chic	Favor
Envy	Dark Yellow Paragon	Circus	Glowing Mandalay
Free Spirit	Eureka	Coral Charm	Lucido
Karma	Gold Champ	Dark Circus	Mandarin
Kiss	Hopscotch	Dark Pomona	Redding
Mountain Snow	Indio	Deep Luv	Red Torch
Parigon	Iridon	Engarde	Salmon Charm
Puritan	Miramar	Loyalty	Salmon Splendor
Raya	Mountain Peak	Luv	Sequest
Solo	Pico	Pasadena	Stoplight
Spirit	Songster	Pomona	Theme
Surf	Spice	Royal Trophy	Torch
Windsong	Sunburst Spirit	Skylight	
Winter Carnival	Sunny Mandalay	Splendor	
	Surfine	Tasca	
	Vista	Tempo	
	Yellow Boaldi	Twilight	
	Yellow Envy		
	Yellow Favor		
	Yellow Ovaro		
	Yellow Tan		
	Yellow Torch		

Source: Adapted from Tayama, Harry K., "Flowering Potted Plants for the Interior Plantscape," Ohio Florists' Bulletin 766: 2, 1993.

require little grooming. Except for *Guzmania* they require less-frequent watering (every 7–14 days) than other flowering plants. Overwatering is harmful. To extend flower longevity, *Guzmania* varieties benefit from having water sprinkled over their vases and inflorescences periodically. For the other genera, do not fill the vases with water; rot is encouraged. With proper care, many bromeliads will last 8 weeks or more indoors.

Orchids are another group of long-lived flowering plants finding more use indoors, both in commercial plantings and at home. The best orchids for low light are species and cultivars of *Paphiopedilum* and *Phalaenopsis* with *Dendrobium* better adapted for moderate- to bright-light areas. *Phalaenposis* have the greatest longevity (up to 3 months) followed by *Dendrobium* and *Paphiopedilum*.

Other flowering plants with long life are *Aphelandra*, *Anthurium*, *Spathiphyllum*, and *Begonia semperflorens culturum* (Wax Begonia).

Figure 12.56 *Kalanchoe.*

Ornamental pepper, *Kalanchoe* (Figure 12.56), 'Rieger' *Begonia*, azalea, and *Cyclamen* remain attractive for 2 to 3 weeks, while *Sinningia*, *Hydrangea*, Easter and hybrid lilies, Amaryllis, *Calceolaria*, and *Cineraria* last 1 to 2 weeks. Bulbous plants such as tulips, crocus, daffodils, and hyacinths last only a few days.

Members of the gesneriad family offer extended enjoyment in interior plantings. *Exacum* and African violet in small pots are ideally suited for tabletops. *Episcia*, *Aeschynanthus*, and *Exacum* make attractive hanging planters.

Improved growing techniques, better shipping methods, and increased grower cooperation have made flowering potted plants readily available, some year-round, others seasonally. Table 12.3 lists availability.

In addition to flowering plants, foliage plants with colored leaves may also be used to add color to a planting of green plants. Croton and *Caladium* have been used effectively. In some situations, the possibility of using flower arrangements should be investigated.

PRESERVED AND ARTIFICIAL PLANTS

Tropical and subtropical foliage plants used indoors cannot be matched for the benefits they provide. Sometimes, however, improper light levels (too low or too high), unsuitable temperatures (too high or too low), inaccessibility, and plant weight may hamper the use of live plants in building interiors. A poor environment will necessitate frequent replacement and additional cost for the client. If a suitable environment for live plants does not exist or cannot be created at a reasonable cost, an alternative is the use of preserved or artificial plants. Preserved plants, mostly palms,

TABLE 12.3   Availability of Flowering Potted Plants

Plant	Availability
Anthurium	January to December
Azalea	September to May
Bromeliads	January to December
Caladium	May to June
Dendranthema (Chrysanthemum)	January to December
Cineraria	March to April
Cyclamen	October to February
Easter Lilies	Two weeks before Easter
Exacum	May to October
Gerbera	March to September
Gloxinia	March to December
Heliconia	June to September
Hyacinth	February to March
Hydrangea	March to May
Kalanchoe	January to December
Narcissus (Daffodil)	February to March
Orchids	January to December
Poinsettia	November to December
Primrose	January to April
Regal Geranium	March to May
Rieger Begonia	January to December
Saintpaulia (African Violet)	January to December
Spathiphyllum	January to December
Streptocarpus	May to June
Tulip	February to April

Source: Adapted from Bennett, Barbara J., "Make Interiorscapes Bloom," Ohio Florists' Association Bulletin 774: 5, 1994; and Harlass, Sherry, "Painting the Interior Landscape," Interior Landscape 9(5): 42, 1992.

are real plants that have been chemically treated to preserve them in their natural state. Artificial plants and foliage are man-made from silk or other materials. Quality man-made plants are very realistic in appearance.

Like it or not, preserved and artificial plants are here to stay. They are part of every interiorscaper's offerings, and their use is increasing. However, they do not, and are not, replacing live material. Clients often present other than environmental reasons for wanting fake plants. One is the perception that they have no maintenance costs. Although they do not require irrigation and grooming, artificial and preserved plants do get dusty, and require regular cleaning just as live plants do. They do not have indefinite life spans, although the frequency of replacement has not been determined. Initial purchase and installation costs of fake plants may exceed that of real plants.

To a public unfamiliar with the cultural requirements of plants indoors, fake plants may send a wrong message. Seeing lush plants in a poor environment, and not realizing that they are not real, may encourage one to install live plants in an unsatisfactory location. When the plant declines or dies quickly, dissatisfaction occurs, and future plant purchases may be curtailed.

As environmentalists, interior plantscapers should use artificial or preserved plants only in conjunction with live plants, and only under conditions where the health of live plants would not be compromised.

The atrium at The Mirage in Las Vegas is one example of the effective use of real and artificial plants. Below and to eye level, live plants, including masses of orchids, create the sense of a South Seas forest. Above eye level, the overstory is artificial plants.

Silk plants, moss, dried flowers, and baskets could be classified as combustible decorative materials and may be regulated. Before use, they should be treated with an approved flame-retardant solution. Interiorscapers must know the potential fire hazards and the laws and penalties involved if a fire should start and property damage or personal injury occur.

NOVELTIES

Interiorscapers are always seeking new plants and new ways to use the tried-and-true workhorses. Recently, old plants in new forms have become available. They offer designers opportunities to create fresh, innovative, exciting plantings. Previously available only on outdoor plants, novel foliage plants are marketed today as standards; braids, ropes, and twists; open-weaves; poodle cuts; bonsai; and other forms.

Standards (Figure 12.57) are normal tree forms with single, clear trunks and shaped heads. *Schefflera actinophylla* 'Amate', *S. arboricola* (Dwarf schefflera), *Ficus benjamina* (weeping fig), *F. maclellandi* 'Alii', and other plants are available as standards in addition to the traditional forms.

Braids (Figure 12.58), ropes, and twists have stems of two or more plants twisted into a braid-like configuration. Eventually, the stems fuse together (graft) where they touch. Open-weaves are similar except that they have open spaces between the stems to prevent fusing. A see-through look is the result. Look for *S. arboricola*, *Dracaena marginata*, *F. benjamina*, *F. maclellandi* 'Alii', and *F. retusa* 'Nitida' in these unusual forms.

Poodle cuts are created by pruning the plants to leave rounded shapes along main stems and sometimes even on branches. Ficus species are especially adapted for training as poodle cuts.

Bonsai are miniature trees and shrubs grown in small containers and possessing all the characteristics of a mature specimen growing in a normal environment. Tropical and some temperate species grown as bonsai are adapted to interior use. Used as specimens on tables and desks, bonsai attract considerable interest and attention.

Figure 12.57 *Ficus benjamina* grown as a standard.

(a)

(b)

Figure 12.58 (a) Braided *Ficus benjamina;* (b) close-up of braid.

PURCHASING PLANTS

Retailers and Plantscapers

The first step in procuring high-quality plants is to get to know the producers. Visit the production areas in Florida, Texas, and California as well as the growers and suppliers in your local area. Deal only with reliable growers of high-quality plants. Plantscapers may want to hand-pick the plants they need.

For foliage plants, quality consists of the sum of several components, including, in decreasing order of importance, condition, cleanliness, size, color, and form. Condition is a rather nebulous term referring to the overall appearance of the plant. Does it appear healthy and fresh and free from injury, pests, and diseases? Are the leaves clean and free of blemishes and pesticide residues? Are the size and form appropriate for the intended use? Is the color normal? Quality assessment is primarily subjective and will vary among individuals.

When purchasing, learn all you can about the cultural practices used in production, because quality begins with the grower. At what light intensities were the plants grown? What was the fertilizer program? If Osmocote or other slow-release fertilizer was used, when was it last applied? What was the pesticide program? Such information will enable the retailer, wholesaler, or plantscaper to properly handle the plants when they arrive. Slow-release fertilizer may have to be removed, repotting may be required if the medium is not satisfactory, and an extended period of acclimatization may be necessary.

Upon delivery, whether to shop, greenhouse, or installation site, the plants should be unpacked immediately. Do not delay. In many instances, plants may have been in dark transit for several days, and it is important to get them immediately under the proper light intensities. In addition, in closed cartons in closed trucks, ethylene buildup may occur and cause plant injury. Ventilation is important. Isolate the newly arrived plants from others. Water them thoroughly and subsequently maintain a substrate moisture content appropriate to each particular kind of plant. Examine the plants carefully for insects, diseases, and injury, and correct any problems that may exist. A temperature of 60 to 65°F (15 to 18°C) is desirable at night, with 70 to 75°F (21 to 24°C) during the day. Temperatures above 75°F (24°C) are not recommended. Following a period of acclimatization, the properly labeled plants may be placed on sale or moved into the plantscape. One further note: plants that leave a shop or greenhouse should be well established and wrapped, particularly in cold weather, to protect them from injury.

Indoor Gardeners

Before setting out to purchase foliage plants for your home, assess the environment in which the plants are to be placed. As previously described, plants may be chosen which are adaptable to the given situation, or the environment may be changed to suit other types of plants. Make a list of plants that will suit both your taste and the environment. Consider such design aspects as shape, size, color, and texture. A knowledgeable retailer can be of great assistance in recommending suitable plants.

Any place that sells healthy plants is a good place to buy, and one should explore all possibilities. A shop or display that is clean and well organized with knowledgeable personnel is the best bet. Good-quality plant material, well displayed and maintained, is a good indication of a well-run business. Mail-order purchases may not be desirable unless selection is limited in your local area. Buy only from reliable, reputable mail-order firms. Consult your local Better Business Bureau if in doubt.

Browse through the shop and look over the entire selection of the plants you seek. Examine them for insects and diseases. Are the leaves healthy and well-formed, or are they misshapen and brown, with many having fallen from the plants? Have the plants been watered? What is the health of the root system? Purchase only vigorous, healthy plants. To do less will result is dissatisfaction.

SUMMARY

More than 300 species of foliage plants are grown commercially, but the list of proven species for maximum satisfaction in interior plantscapes is much smaller. Tropical and subtropical species are more suitable than those of temperate regions.

Flowering plants are used to add color to interior plantscapes. Maintenance is more demanding, requiring frequent replacement, more frequent watering, and brighter light; hence, costs are high. Colored-leaved plants add color and the use of flower arrangements should be considered. Use of artificial and preserved plants is increasing. Novelty forms including standards, braids, ropes, twists, and bonsai are also available.

Retailers and plantscapers should get to know the producers and purchase from reliable growers of high-quality plants. To properly handle plants, learn the cultural practices of the grower. Upon delivery, open cartons, provide light and a suitable temperature, and water the plants. Examine the plants for insects and diseases, and acclimatize before sale or installation, if necessary.

Home gardeners should select plants to suit their taste and the environment. Buy from a reliable, knowledgeable retailer who consistently sells high-quality, healthy plants.

REFERENCES

ALCA *Guide to Interior Landscape Specifications*, 4th ed., Associated Landscape Contractors of America, 12200 Sunrise Valley Drive, Suite 150, Reston, Va. 22091, 1988.

AULENBACH, B.: "Color Spots and Change Out Plants for Interiorscapes," *Ohio Florists' Association Bulletin* 649: 3–5, 1983.

BENNETT, B. J.: "Make Interiorscapes Bloom," *Ohio Florists' Association Bulletin* 774: 5, 1994.

BICKFORD, E. D.: "Choose the Best Light Source," *Florists' Review* 161(4161): 29ff., 1977.

BIZON, K. J.: "Maintaining Blooming Plants," *Interior Landscape* 9(5): 52–53, 1992.

BOODLEY, J.: "Local Handling," Lecture, 1978 National Tropical Foliage Short Course, Orlando, Fla.

BURCH, D. G.: "Palm Reading," *Interior Landscape Industry* 1(2): 35–37, 39–40, 1984.

CATHEY, H. M.: "Recommended Light Levels for Selected Decorative Indoor Plants," *American Horticulturist* 55(1): 39, 1976.

CHATHAM, C.: "Maintenance Guidelines for Interior Flowers," *Interior Landscape Industry* 1(5): 34–35, 1984.

CIALONE, J.: "Color—A Key Factor in Your Interiorscape Plan," *Interscape* 6(42): 21–24, 1984.

_____: "Everything Old is New Again," *Interior Landscape Industry* 4(6): 50–52, 1987.

CLEIN, M.: "Survey Results Gave Mall of America Designer Direction," *Interiorscape* 12(6): 58–60, 1993.

_____: "Frank Smith: It's an Orchid World After All!," *Interiorscape* 12(1): 46–52, 54, 1993.

"Color for Interiors," *Grower Talks* 47(5): 26–28, 30, 1983.

CONKLIN, E.: "Interior Plantings Bring Nature Indoors," *American Nurseryman* 140(2): 12ff., 1974.

CONOVER, C. A., T. J. SHEEHAN, and D. B. MCCONNELL: *Using Florida Grown Foliage Plants*, Bulletin 746, Florida Agricultural Experiment Station, University of Florida, Gainesville, Fla., 1971.

COPLEY, K.: "Plants for Medium Light Situations," *Grounds Maintenance* 13(2): 68, 72, 74, 1978.

DE NEVE, R. T.: "Alternative Interior Plants," *Interior Landscape Industry* 1(2): 24–26, 28–33, 1984.

DEWERTH, A. F.: *Indoor Landscaping with Living Foliage Plants*, B-1118, Texas A & M University, The Texas Agricultural Experiment Station, College Station, Tex., 1972.

DONSELMAN, H.: "A Preponderance of Palms," *Interior Landscape Industry* 6(10): 56–59, 61–64, 1989.

FREEMAN, R. N.: "Moisture, Light and Temperature Needs of Some Tropical Foliage Plants: Cornell Foliage and Epiphytic Mixes," *Florists' Review* 160(4156): 32, 1977.

FRIENDSHIP, R. A., M. J. TSUJITA, and D. P. ORMROD: "Temperate Woody Species as Interior Landscape Plants," *Journal of Environmental Horticulture* 4(2): 47–51, 1986.

FURUTA, T.: *Plants Indoors: Selections for Various Environmental Conditions*, Leaflet 2898, Division of Agricultural Sciences, University of California, Berkeley, Calif., 1976.

GRAF, A. B.: *Exotic Plant Manual*, 2nd ed., Roehrs Company, East Rutherford, N.J., 1972.

HAMMER, N.: "Using Preserved," *Interiorscape* 12(1): 56, 58–59, 1993.

HARBAUGH, B. K.: "Flowering Plants for Interiors," *Foliage Digest* 7(10): 4–7, 1984.

HARLASS, S.: "Painting the Interior Landscape," *Interior Landscape* 9(5): 38–43, 1992.

HENLEY, R. W., and D. B. MCCONNELL: "Plant Nomenclature," *Foliage Digest* 9(6): 35–38, 1983.

KEEVER, G. J. and G. S. COBB: "Temperate Zone Woody Plants for Interior Environments," *Journal of Environmental Horticulture* 4(1): 16–18, 1986.

_____, and J. C. STEPHENSON: "Interior Performance of Temperate Zone Landscape Plants," *Interiorscape* 9(6): 6–7, 53, 1990.

KORSTAD, D. A.: "Selecting Plants for Interior Plantscaping," *Florists' Review* 171(4426): 34, 37–38, 40, 1982.

MARCHANT, B.: "Blooming Beauties," *Interior Landscape Industry* 1(5): 27–32, 1984.

_____: "Interior Landscape Renovation—A Case Study," *Interior Landscape Industry* 2(3): 50–58, 1985.

MCCONNELL, D. B., and C. A. CONOVER: "Commonly Available Foliage Plants Best Adapted to Interior Environments," *Proceedings of the Florida State Horticultural Society* 86: 478–479, 1973.

MOREY, C.: "Safety, Liability, or Law," *Interiorscape* 9(5): 35, 59, 1990.

NELL, T. A. and J. E. BARRETT: *Longevity of Flowering Potted Plants in the Interiorscape*, Staff Report Series ORH 86–3, Department of Ornamental Horticulture, University of Florida, Gainesville.

NEWMAN, L.: "The New Workhorses," *Interior Landscape* 9(5): 44–50, 1992.

OTT, R.: "A Plethora of Palms," *Interior Landscape Industry* 9(3): 32–35, 1992.

PEPPLER, K. Z.: "Mum's the Word," *Interior Landscape Industry* 8(8): 18–23, 1991.

_____: "Shades of Success," *Interior Landscape Industry* 8(5): 22–24, 1991.

Plants for Interior Landscapes, Associated Landscape Contractors of America, 405 N. Washington Street, Falls Church, Va., 22046, 1981.

PRINCE, T. A. and T. L. PRINCE: "How Many Are Saying It with Flowers?" *Ohio Florists' Association Bulletin* 670: 1–4, 1985.

"Small Plant Materials," *Interscape* 6(3): 1–4, 1984.

STABY, G. L., J. L. ROBERTSON, D. C. KIPLINGER, and C. A. CONOVER: *Chain of Life*, Horticultural Series No. 432 (1978), Ohio Florists Association, 2001 Fyffe Court, Columbus, Ohio 43210.

TAYAMA, H. K.: "Flowering Potted Plants for the Interior Plantscape," *Ohio Florists' Bulletin* 766: 1–3, 1993.

THAMES, G., and M. R. HARRISON: *Foliage Plants for Interiors*, Extension Bulletin 327-A, Cooperative Extension Service, Cook College, Rutgers University, New Brunswick, N.J., 1966.

Installation and Maintenance Contracts

The success of an interior plantscape depends on the cooperation of a team of experts, which may include the architect, landscape architect, interior designer, engineers, growers, and horticulturists of the plantscape firm. Horticultural knowledge plus common sense will enable the interior plantscape contractor to make a valuable contribution to the design, installation, and maintenance decisions concerning proposed plantings. Contracts will protect both the contractor and client, and prevent misunderstandings.

INTERIOR LANDSCAPE FIRMS

Firms involved in the profession of interior landscaping may be found from coast to coast. While many firms are the outgrowth of other established businesses, such as landscape contracting, nursery, design, and retail flower shops, most began as small operations to meet the sudden demands of the seventies. Although some operate retail and/or wholesale outlets, the major interior landscape firms today are interiorscapers only. There are hundreds of businesses nationwide ranging in size from small firms with a few employees and gross volume under $50,000 to large corporations with hundreds of workers and annual volumes exceeding $83 million.

As a group, interiorscapers sell a customer installed plants and guaranteed maintenance. This arrangement accounts for the majority of sales, with the balance coming from plant leases and rentals. Large firms derive a greater portion of their income from leases than do smaller companies, while the reverse is true for plant rentals. Generally speaking, maintenance provides the largest share of income to the interior plantscape business. Most firms do in-house design work.

Interior landscape firms handle all kinds of accounts, from hotels, restaurants, and shopping malls to offices, showrooms, retail stores, institutions, and residences. Offices appear to top the list, while residences account for a small percentage of the business.

Interior landscape contractors rely on design professionals such as architects and interior designers for a considerable portion of their new business. Repeat business accounts for one-fourth to one-half of their annual volume.

Various methods are used in conducting plantscape businesses, and successful firms may be involved in more than one approach. When outright sale is used, the plant material is the property of the client. Once installed, the contractor is not concerned with aftercare, although many guarantee plants for 30 to 60 days. Plant rentals are made for short or long terms, with ownership of the plants remaining with the contractor.

Three other modes of operation are similar: install and maintain; install, maintain, and guarantee; and lease and maintain. In the first two, the client owns the plants. Replacement procedures depend upon the terms of the guarantee. In the last situation, the contractor owns the plants and is expected to keep the planting in tip-top shape at all times. For the plantscaper, there is a preference for outright sales. Leasing ties up capital and reduces cash flow.

INSTALLATION GUIDELINES

Most contractors prefer to negotiate installation contracts. However, this method is not always possible and frequently one must get involved in competitive bidding. When one has been invited to submit a bid for a proposed interior plantscape, careful attention must be given to the specifications, as well as the physical environment, in order to determine the costs involved. Not to do so could result in major, and often expensive, problems later. All costs must be anticipated and covered in the bid.

Start with a set of blueprints and specifications. Specifications are the written directions for the project; the blueprints or working drawings, the illustrations. Depending upon the complexity of the interior landscape, working drawings may include layout, grading, planting, and irrigation plans plus details and a plant list. Small, simple projects may require only a single drawing.

A layout plan provides dimensions for all of the elements of the design and locates and identifies everything, except plants, that the contractor is responsible for. A grading plan is needed when the interiorscape contains planting beds or other areas where the surfaces are not flat. The planting plan shows the locations, types, numbers, and sometimes the sizes of the plants to be used, and is perhaps the most important drawing. If an irrigation system is to be installed, an irrigation plan is provided. Details are specialized drawings showing execution of complex features of the landscape. Crosssections of plantings as viewed from bottom to top are frequently used details. The plant list is integral to the final document and should include common names, scientific names, quantities specified, and plant sizes. It should be included on the planting plan, if possible, and may also be incorporated into the specifications.

If the facility exists, visit the site and perform an interior site analysis. Learn the total design concept. What kinds of furnishings are to be used and what are the colors of walls, screens, carpets, and upholstery? Know how to read the furniture plan. The plan will show where each plant is to be located. Look at the lighting. Is it adequate? Will it be task lighting rather than general, with low intensities in the areas where plants are frequently located? Will the lights be turned off at 5:00 P.M.? If natural light is to be used, will there be obstructions from ledges, blinds, or drapes? Is the glass bronze or clear?

Learn the heating and cooling specifications of the areas to be planted to determine the temperature range. Are the heat and air conditioning turned off on weekends and holidays? If so, the quality of the plants will decline more rapidly than in a controlled environment.

Are there any problems with accessibility to the site? How big are the doors? Are the elevators accessible, and how many can be used? Are they large enough for large plants? If not, a determination must be made as to how large plants will be moved to the site. Will other persons be using the elevators during the delivery of the plants? Is it possible for the truck to be unloaded in an enclosed area?

Consider labor requirements. How many person-hours will be involved? Is there a labor union factor that must be considered in your estimate?

Work closely with the architect, interior designer, or client. After a careful analysis of the site, you may want to suggest changes in the specified plant material or its location. Recommend species best suited for the particular environment and which will achieve the overall design effect. If the choice of plants is left to the plantscaper, a knowledge of the budget available is essential.

The owner, or owner's agent, may want to approve plants at the grower's site, particularly if they are large, expensive specimens. If so, add travel expenses to your costs. It is desirable to present a proposal to the client itemizing each plant as to location, name, and size. If appropriate, the type of container should also be listed. The price should be presented as the total costs for plants, labor, containers, and other items, rather than item by item. This will prevent the client, who may have no understanding of design, use, or suitability, from trading off more expensive plants for ones that are cheaper but less satisfactory.

When will the planting be installed? What is the starting date? Is there a placement schedule? Because of extra protection required, installation is more costly in the winter than at other times. How much lead time do you have? You must have time to confirm orders and acquire and acclimatize the plants. Working with growers, determine the conditioning requirements and prepare the necessary facility. Do you have suitable equipment to handle the job? If not, can it be rented? Are there weight restrictions?

If there is a postponement in the installation caused by delays in construction, what protection does the interior plantscape contractor have against the extra costs incurred? What penalties will be charged should the plantscaper not complete the work on time?

The type of growing medium should be specified, particularly where weight may be a factor. If the installer will handle the ultimate maintenance, he or she should have considerable input into the choice of medium, as this will affect future practices. If no maintenance is involved, the owner or designer should specify the growing medium.

What provisions will be made for storage of equipment, media, and plants? Security is essential for all stored materials, and heat, light, and water are imperative for the plants.

All of the financial requirements for the project should be determined in advance. Do you have adequate funds, or can they be procured to meet your expenses until payments are received from your client? What are the terms of payment? Do you have a performance bond? Another financial consideration is insurance. Adequate liability, property damage, automobile, Workers' Compensation, and other insurance coverage should be in effect.

Before plants are installed, all construction overhead and in the vicinity of the planting should be completed and all debris removed. Heavy furniture and floor coverings should be in place. Heating, ventilation, and air-conditioning systems should be operating and balanced; lighting systems should be fully active. Dust and fumes should be minimal. Hot- and cold-water mixing faucets should have been installed and be operating.

Plants are handled using dollies and other types of carts. Large, heavy specimens may be moved with front-end loaders or portable hoists (Figure 13.1). Grow-

(a)   (b)

Figure 13.1 (a) Hand-operated hydraulic life facilitates moving planters and other heavy objects: (b) front-end loaders, or other heavy equipment, may be required on some jobs. Weight restrictions should be investigated.

ing media, mulch, and stone are easily handled in bags or bales. Do not water the plants just prior to installation, as they will be lighter and easier to handle. A thorough watering following placement is essential.

Until the completed installation is accepted by the owner, he or she should notify all others not to water or care for the plants, as this is the responsibility of the contractor. Use of the planters for trash or cigarettes should be discouraged.

Terms of payment, and a date when the quote expires, should be stipulated.

Once completed, the bid package for installation and/or maintenance is submitted to the client. This is an excellent opportunity to make a favorable impression, so make a high-quality presentation. Be sure that the bid sheet is perfect with no erasures, written-in corrections, or grammatical errors. Here, and in the cover letter, there should be no misspellings, especially the name of the client and that of the firm. Any brochures or pamphlets should be professionally printed; project drawings should be neat, legible, and complete and project a positive image of your company. To do less could mean the loss of the job.

To assist plantscapers in the preparation of bids and contracts, the Associated Landscape Contractors of America has prepared *A Guide to Interior Landscaping*. A copy may be purchased from the ALCA, 12200 Sunrise Valley Drive, Suite 150, Reston, VA 22091. Legal counsel is recommended in the preparation of any contract.

MAINTENANCE

Foliage plants in the indoor garden must be healthy, clean, and attractive at all times. A low loss rate, healthy plants, and client satisfaction should be the primary goals of every maintenance program. For maximum satisfaction, the firm responsible for the initial installation, or its agent, should maintain the plantscape for at least the first year.

Maintenance is usually on a guaranteed basis with replacement at no cost to the client. Plants under contracted maintenance will usually receive better care than those left to in-house management. Trained technicians maintain the plantscape rather than possibly inexperienced individuals whose regular job demands time and who are not trained to detect plant problems at an early stage. Management personnel must supervise maintenance people, and they may not have the knowledge to do so. When people are sick or on vacation, the plants may go untended. All factors considered, the time required to maintain interior plants may not be spent and there may be a rapid decline in appearance of the plantings.

By starting with high-quality plants that have been properly acclimatized, installed, and groomed and using knowledgeable, trained technicians, maintenance becomes routine. A successful firm should strive for a loss (replacement) of less than 10% per year.

Many firms will not maintain plants they have not installed. Those that do frequently require that the planting be brought up to their standards. Since the new firm had no control over previous practices, guarantees during the first several months may be limited.

Personnel

Maintenance technicians are key personnel in the success of the plantscape. Individuals with a deep interest in plants and plant health, and who are eager to learn, usually make good maintainers. Several years of greenhouse or landscape experience is recommended, and either an Associate or Baccalaureate degree in horticulture or botany is desirable. Experience with another maintenance firm may be worthless, or treacherous, if the new employee brings bad habits to the new firm. The work involves considerable contact with others, so a pleasant demeanor is an asset. Since there is little direct supervision, competency and self-motivation are critical.

Regardless of the background and experience of maintainers, a training period is essential. A program takes several weeks and involves classroom work as well as on-the-job training with an experienced technician. Maintainers must not only know about plants and plant maintenance, they must understand your business: the contractual arrangements, the paperwork, and the client relationships.

Every effort must be made to minimize employee turnover. Training is costly, but perhaps more important, it dispels the uneasiness that may develop in clients when they see a steady stream of new people. Pay scales competitive with other occupations, opportunities for advancement, and employee-benefit programs are some of the ways employers have of retaining employees.

Procedure

Regular servicing once a week, or more often, is the norm, as routine evaluation and care are the key to maintenance of high-quality plants. Assign maintenance technicians to routes so that care is continuous and the clients become familiar with the personnel. There should be no surprises.

Provide your maintainers with a van or station wagon, uniforms, and the proper tools. Simple tools and equipment that might be used every day include watering cans (2-gal size), half apron with side loop, sponges and paper towels, pruning shears, folding saw, misting spray bottle, liquid soap such as Ivory, soil probe, hand lens, plastic garbage bags, and dust pan and brush. The technicians should also have available in their vehicles a hose and spray attachment, water wands, breakers, and shut-offs; a watering machine; pesticides labeled for interior use, protective clothing, and measuring equipment (spoons, cups, and scale); a light meter; and a pH/salinity meter. A variety of other specialty equipment may be needed to perform unusual or infrequent tasks. Depending upon the frequency of use and investment, this equipment may be either owned or rented. Compared with outdoors, interior landscape maintenance is labor-intensive. Finally, have adequate liability insurance coverage for injury to personnel as well as to furniture, carpets, desks, walls, drapes, and so on, belonging to clients.

Perform all maintenance during the day so that clients are aware that it is being done. Spraying, when necessary, may have to be done at night. Have a client representative sign for all work performed, and let clients know monthly the services they have purchased.

Information Report 76-3, published by Texas A & M University, is a case study of the costs of planting and maintaining the plantings in several buildings at Texas A & M University, and provides useful guidelines for consumers and plantscapers.

Planting is usually a one-time operation lasting over a year. Rotation—moving the plant from one location to another or to a greenhouse—turning to maintain straight growth and alignment, loss to insect or disease injury, and pilferage may necessitate replacement and increased planting costs. Pilferage of whole plants and plant parts (cuttings) is a common practice. According to the report, planting costs include growing medium, plant, container, and labor. Table 13.1 shows the costs involved.

Potting in the greenhouse was less of an operation than potting at the site, because only the potted plants had to be moved, not all of the equipment and supplies needed. Delivery, including loading, transportation time, unloading and moving to site, and return time, for a plant potted at the greenhouse was less than half that for on-site potting.

Maintenance was a regular activity; some jobs were performed on a daily basis, others weekly or monthly, and some three or four times a year (Table 13.2). One worker per installation was usually enough, with plant replacement requiring assistance from others.

Watering is the most repetitive maintenance task. Frequency of watering will vary with the size and kind of plant as well as the season and may be reduced with the use of subirrigation planters. During the heating season, plants will require more frequent watering. In the Texas study, almost 25% of the cost of maintenance was for watering.

To maintain the aesthetics of the landscape, the plants must be cleaned regularly. Wiping each leaf is a time-consuming operation. Plants less than 4 ft tall, which do not require the use of a ladder, will require 30 to 45 min to clean if they have a large number of leaves, and 15 to 30 min for a small number of leaves. Plants over 4 ft tall requiring the use of a ladder take 3 hrs and 1 hr for large and small numbers of leaves, respectively. Other maintenance operations are listed in the table.

For ease of maintenance and replacement, an inventory of every planting should be maintained and a worksheet kept for each plant (Figure 13.2). This information is intended as a guideline only, and every individual involved in installation and maintenance must consider the costs as they relate to one's own business and each specific job.

TABLE 13.1 Costs of Planting

Container	(1) Container Cost	(2) Average Cost/ Plant	(3) Growing Medium (ft^2)	(4) Cost of Medium	(5) Planting Time/Plant (hr)	(6) Labor Cost[a]	Cost of Planting Individual Plants: Columns (1) + (2) + (4) + (6)
18-in. copper	$300.00	$ 50.00	3.53	$ 7.94	1	11.00	$368.94
24-in. metal	50.00	75.00	8.00	18.00	1	11.00	154.00
24-in. fiberglass	95.00	90.00	6.28	14.13	1	11.00	210.13
24-in. redwood	250.00	125.00	8.00	18.00	1	11.00	404.00
42-in. redwood	375.00	175.00	24.50	55.13	1	11.00	616.13
48-in. redwood	500.00	200.00	32.00	72.00	1	11.00	783.00
10-in. clay	1.95	15.00	1.00	2.25	1	11.00	30.20

[a]Two workers, one at $5.00/hr, one at $6.00/hr, not including Social Security, Workers' Compensation, fringe benefits, and equipment costs.

Source: Adapted from H. B. Sorensen, W. L. Vitopil, and D. Leighman, Costs of Planting and Maintaining Ornamental Plants in Public Buildings, TAMU, 1975: A Case Study, D.I.R. 76–3, Texas A & M University, College Station, Tex., 1976, p. 21.

TABLE 13.2 Maintenance Operations for a Large Containerized Plant

Operation	(1) Time Between Operations	(2) Times Performed Per Year	(3) Time for Each Operation (min)	(4) Total Time (Min): Cols. (2) × (3)	(5) Number of Hours	(6) Labor Cost/hr[a]	(7) Cost Per Plant: Cols. (5) × (6)
Watering—air conditioning	10 days	24	7	168	2.8	$ 5.00	$14.00
Watering—heat	5 days	24	7	168	2.8	5.00	14.00
Watering—additional	As needed	24	7	168	2.8	5.00	14.00
Trimming	3/yr	3	4	12	0.2	5.00	1.00
Pest Control	1/month	12	3	36	0.6	5.00	3.00
Fertilizing	2/yr	2	2	4	0.07	5.00	0.35
Cleanup	10 days	36	2	72	1.2	5.00	6.00
Clean plants	3/yr	3	60	180	3.0	5.00	15.00
Foreman	2/month	24	5	120	2.0	15.00	30.00
Total							$97.35

[a]Not including Social Security, Workers' Compensation, fringe benefits, and equipment costs.

Source: Adapted from H. B. Sorensen, W. L. Vitopil, and D. Leighman, *Costs of Planting and Maintaining Ornamental Plants in Public Buildings*, *TAMU, 1975: A Case Study*, D.I.R. 76–3, Texas A & M University, College Station, Tex., 1976. p. 25.

Scientific name:								Common name:								
Source:																
Date planted:							Container number:				Location:					
Type container:								Type medium:								

Date	Watered	Rotated	Pest control	Pruned	Cleaned	Fertilized	Replaced		Date	Watered	Rotated	Pest control	Pruned	Cleaned	Fertilized	Replaced	

Figure 13.2 Worksheet for the maintenance of interior plantscapes.

Contracts

A maintenance contract is desirable to protect both the contractor and the client and eliminate misunderstandings. Everything should be put in writing.

Among the items for consideration when preparing a maintenance contract are:

1. Frequency and type of maintenance.
2. Environmental considerations.
3. Replacement.
4. Guarantee restrictions.
5. Cancellation rights.
6. Fees.
7. Insurance.

Frequency and Type of Maintenance. Maintenance should be at least once a week and normally includes inspection, watering, dusting, and cleaning. Other maintenance, such as fertilizing and spraying, is performed as required. The contractor should have access to all plants during normal working hours, as well as access to hot- and cold-running water. If required, a secured storage space for maintenance equipment should be provided.

Environmental Considerations. The plants have been selected for and placed in a particular environmental situation. Changes in the environment over which the maintenance firm has no control may injure the plants. The contract should stipulate the minimum and maximum temperatures, and the intensity of the light and hours of illumination. The contractor should not be responsible for loss or damage caused by deviations from the specified ranges. If a client moves a plant without permission, the contractor should not be responsible.

Replacement. A replacement clause should be part of any maintenance agreement. The plants represent the contractor and should always look their best. When plants are rented or leased, they should be presentable at all times and the client should have no reason to complain. With client-owned plants, if replacement is not automatic, there is a tendency to try to extend the life of the plant, and deteriorating plants remain in the plantscape. Replacement should be automatic and be done during working hours by a well-trained, neat, and courteous crew. An inventory of the plantscape by location will facilitate changing the plants rapidly and economically. The determination that a plant needs to be changed should be at the discretion of the service technician. The new plant should be of the same size specified in the original contract; however, the right to substitute plant type should be permitted. Depending on the contractor, the removed plant may be taken to a greenhouse or other area for rejuvenation, placed in a more favorable location in the plantscape, or destroyed. Since replacement is immediate, the contractor must maintain an inventory of acclimatized plants.

Guarantees. Guarantees usually consist of assurance that high-quality plants will be kept in specific locations during the period of the contract. The contractor must be protected against loss from factors over which he or she has no control. The guarantee should be void and the client responsible for the full cost of the plant for damage due to unauthorized watering or other care by the client firm, its employees, or cleaning contractor; accidental or malicious acts; fire, theft, cold, cleaning chemicals, and extended loss of light or heat due to power failure. Employers should inform their employees of the contract and discourage "assistance" in caring for the plants.

Cancellation Rights. The agreement should be renewed automatically unless written notification 30 days prior to the annual termination date is given by either party. Any charge associated with cancellation should be stipulated. The right to renegotiate maintenance fees annually should be stipulated.

Fees. Agreements are usually for a period of 1 year, with fees based upon a weekly or monthly schedule, billed and payable monthly or quarterly, in advance. In determining the service fee, all costs of doing business must be considered. Labor costs, including salary, Social Security taxes, and the cost of fringe benefits, are an important factor, and may be calculated as shown in Table 13.2.

Other costs which must be retrieved are the initial cost of the plants; the replacement cost of plants and the labor involved; travel to and from the site; the cost of supplies and materials; overhead costs, such as management salaries, billing, utilities, rent, insurance, and taxes; and the profit margin. Failure to consider all costs involved will result in an understatement of the fee and will cost the contractor money. Customized software is available for calculating the service fee. Quality programs minimize the possibility of neglecting contractor costs and facilitate accurate pricing. Since each interior plantscape is unique, fees may vary from one client to another. Payment terms should be stipulated. A reasonable method is net 10 days with a 1% per month service charge for accounts 30 days overdue. If payment is not received in 60 days, the contractor should reserve the right to enter the premises to recover leased plants and containers. A good business practice would be a monthly discussion with each client of the services they are buying. This establishes rapport between the contractor and the client, and fosters a partnership in the concern for the well-being of the living decor.

Insurance. The contractor must carry full liability coverage.

SUMMARY

A successful interior plantscape depends upon the cooperation of a team of experts. A new profession has evolved, that of the interior plantscaper, whose firm provides design, sales and rentals, installation, and maintenance.

Contracts for both installation and/or maintenance are desirable to protect client and contractor and prevent misunderstandings. In bidding, consideration must be given to the physical environment (light, heat, cold, ventilation, water), accessibility of site, labor needs, plant acquisition, acclimatization, season of year, time framework, equipment, growing media, postponements, storage, and financing.

Plants must be healthy, clean, and attractive at all times. The installer or agent should maintain the plantscape for at least the first year. Maintenance personnel should be interested in plants and plant health and be well trained. Regular servicing includes watering, cleaning, dusting, and trimming plus other maintenance as required. Contracts are desirable and should include frequency and type of maintenance, environmental considerations, replacement, guarantee restrictions, cancellation rights, and fees.

REFERENCES

BROWNE, M.: "Bidding Basics," *Interior Landscape* 12(2): 14–17, 1995.

DICKMAN, S. D., H. FEINSTEIN, and J. ALMEIDA: "Establishing an Interior Landscape Firm," *Southern Florist and Nurseryman* 94(42): 22ff., 1982.

Guide to Interior Landscape Specifications, Associated Landscape Contractors of America, 12200 Sunrise Valley Drive, Suite 150, Reston, Va. 22091, 1988.

HAMMER, N. R.: "Bidding Genius," *Interior Landscape Industry* 6(1): 84–86, 1989.

____: "Creating Contract Documents," *Interior Landscape Industry* 8(6): 32–39, 1991.

HOROWITZ, C.: "The Planning, Scheduling and Delivery of Materials for the Large Installation," *Interscape* 4(21): 8–9, 12, 1982.

KORSTAD, F. R.: "Finding That Special Person; The Interiorscape Maintenance Tech," *Interiorscape* 2(2): 40, 42, 1983.

KOTEEN, J.: "To Lease or Not to Lease, That Is the Real Question," *Interscape* 5(39): 29–30, 1983.

LEIDER, G. F.: "Plant Rentals and Contracts," *Proceedings: From a Plant's Point of View*, pp. 52–55, Society of American Florists, Washington, D.C., 1975.

MARTELLI, N. N.: "Practical Pricing," *Interior Landscape Industry* 7(12): 36–39, 1990.

MASTICK, D. F.: "Basic Maintenance of Interior Landscapes," *American Nurseryman* 156(10): 63–70, 1982.

____: "Guaranteeing Plant Maintenance—Protect Yourself," *Florists' Review* 162(4193): 60, 104–105, 1978.

NEWMAN, L.: "Equipment for Special Services," *Interior Landscape Industry* 8(12): 36–42, 1991.

____: "Maintenance Equipment Made Easy," *Interior Landscape Industry* 8(11): 40–42, 1991.

NORRIS, P.: "Interiorscapers Should Spell Out Guidelines in Contracts," *Nursery Manager* 1(3); 67ff., 1985.

RAIMONDI, C. S.: "It's Tough Being a Tech," *Interior Landscape* 10(2): 52–56, 1993.

SORENSEN, H. B., W. L. VITOPIL, and D. LEIGHMAN: *Costs of Planting and Maintaining Ornamental Plants in Public Buildings, TAMU, 1975: A Case Study*, D.I.R. 76–3, Department of Agricultural Economics, Texas Agricultural Experiment Station, Texas A & M University, College Station, Tex., 1976.

Glossary

Abscission: Loss, or dropping, of plant parts, especially leaves, by natural separation.

Absorption: Movement of materials into the cells of plants.

Acclimatization: Adaptation of a species from one environment to another.

Aggregate: Secondary soil particle composed of many soil particles bound together.

Anthocyanin: Red, blue, or purple, water-soluble pigment found in the vacuoles of plant cells.

Artificial Light: Illumination from lamps, including incandescent, fluorescent, and high-energy discharge.

Ascocarp: Structure in certain fungi, containing asci or sport-producing sacs.

Atrium: An open courtyard used for planting, which is enclosed on all sides by a building.

Biological Control: Use of natural enemies (beneficials) to reduce insect and mite infestations.

Bulk Density: The weight of a substance per unit volume.

Canker: Dead area, usually on a stem, surrounded by living tissue.

Capillary Water: Water held in the small pores of a growing medium, and usually considered available to plants.

Cation-Exchange Capacity: Sum of all the positively charged ions (cations) that a soil can hold on the surface of soil, clay, or organic particles.

Chilling Injury: Injury associated with low, but not freezing temperature.

Chlorophyll: Green pigment of plants and the absorber of light in photosynthesis.

Chloroplast: Specialized structure in green plant cells which contains the chlorophyll.

Chlorosis: Yellow coloration caused by the absence of chlorophyll due to its failure to develop a result of some physical factor, such as improper light, temperature, or nutrition.

Coalescence: Growing together, forming a larger unit, characteristic of the individual spots or lesions of some plant diseases.

Compensation Point: Light intensity at which the rate of photosynthesis is equal to the rate of respiration.

Complete Fertilizer: Fertilizer containing nitrogen, phosphorus, and potassium.

Conductivity meter: Instrument that measures conductivity of the soil solution. Used to determine soluble-salt levels in the growing medium.

Conservatory: Glass-covered house or room for growing and displaying plants, in which an ideal environment may be maintained.

Container Capacity: Water held in the growing medium of a container following free drainage.

Cultural Control: One of the components of Integrated Pest Management. Pest prevention or control without the use of chemical or biological techniques. Examples include using horticultural practices that minimize plant stress, good housekeeping (sanitation), physical control such as barriers and traps, and resistant cultivars.

Desiccation: Drying of plant tissue.

Dormancy: Period of inactivity during which a plant will not grow, regardless of how favorable the environment.

Double-Potting: Placing a pot containing a plant inside a slightly larger watertight container.

Edema: Abnormal swelling of tissues due to accumulation of fluids, frequently seen as eruptions of the leaf surface.

Emulsifiable Concentrate (EC): Concentrated pesticide or fungicide in liquid form.

Epinasty: Downward bending of petioles in response to disease or atmospheric pollutants.

Ethylene (C_2H_4): Colorless gas produced by plants and hydrocarbon combustion, which is phytotoxic.

Fertilizer: Mixture of minerals providing the essential elements for plant growth.

Fertilizer Analysis: Percentage of nitrogen (N), phosphorus (P_2O_5), and potassium (K_2O) in a fertilizer. 20–20–20 and 10–10–10 are examples.

Fertilizer Ratio: Proportion of nitrogen, phosphorus, and potassium in a fertilizer. A fertilizer analysis of 14–7–7 has a ratio of 2–1–1.

Flowable Pesticide: A suspension of finely ground wettable powder containing water and a wetting agent.

Fluorescent Lamp: Low-pressure mercury discharge lamp in which light is produced by the passage of electricity through a gas enclosed in a tube.

Footcandle: Unit of illumination equivalent to the light produced by a standard candle at a distance of 1 ft.

Grana: Series of stacked membranes within the chloroplast which contain the chlorophyll.

Gravitational Water: Water that drains quickly from large pores due to the force of gravity. It is only slightly available for absorption by roots.

Hardening: *See* Acclimatization.

Hydrophilic Polymer: Synthetic, water-absorbing granule added to a growing medium, capable of absorbing up to 800 times its dry weight in water, for gradual release to plant roots.

Hydroponics: Maintenance of plants in a liquid solution or inert medium supplemented with all the essential elements.

Hygrometer: Instrument for determining relative humidity.

Incandescent Lamp: Lamp whose light is due to a filament that is made to glow by the passage of an electric current through it.

Incipient: Without distinctive symptoms or characteristics.

Indeterminate-Day Plant: Plant that is not affected by day length.

Inorganic Fertilizer: Fertilizer in which the elements are derived from chemical, not biological, sources.

Instar: Insect in any one of its periods of growth between molts.

Integrated Pest Management: Manipulation of all plant management practices to control pests with minimum use of energy and chemicals. Uses cultural (preventive), biological, and chemical methods.

IPM: See Integrated Pest Management.

Jardiniere: Decorative container for holding plants, usually without drainage.

Larva: Immature stage in a life cycle, often restricted to feeding stages, such as caterpillars and maggots.

Leach: Drenching the growing medium with water to wash out excess soluble salts.

Light Intensity: Brightness of light.

Light Quality: Color or spectral-emission discharge (S.E.D.) of a light source.

Loam: Soil which is a combination of sand, silt, and clay and does not exhibit the dominant physical properties of any of the three groups.

Long-Day Plant: Responds when the day length is greater than a certain number of hours.

Lux: International unit of illumination. The light received by a surface at a distance of 1 meter from a light source whose intensity is taken as unity. One lux = 0.09 footcandle.

Macro Nutrients: Elements used by plants in relatively large amounts, including nitrogen, phosphorus, potassium, calcium, magnesium, and sulfur.

Material Safety Data Sheet: A technical report produced by chemical manufacturers and importers which documents the hazards of their chemicals (pesticides, etc.) and communicates them to users. An accessible file of MSDSs must be maintained for all hazardous chemicals used. Request from chemical suppliers.

Mercury Lamp: Lamp producing light by means of an electric arc in mercury vapor.

Mesophyll: Thin-walled living tissue between the epidermal surfaces of leaves, specialized for photosynthesis.

Metal-Halide Lamp: Similar to mercury, but with a different S.E.D. due to metal additives and a rare gas in the lamp.

Micro Nutrients: Elements used by plants in small quantities, including iron, zinc, manganese, boron, copper, molybdenum, and chlorine.

MSDS: *See* Material Safety Data Sheet.

Mulch: Material applied to the surface of the growing medium to prevent excess drying, to prevent rapid temperature change, to prevent weed growth, and for decorative purposes.

Mutant: New type of organism produced by the spontaneous change in the expression of a gene (mutation).

Mycelium: Plant body of a fungus, composed of thread-like filaments called hyphae.

Nanometer: Metric unit of measurement equal to 10^{-9} meter, 10^{-6} mm, 10^{-3} micron, or 10 angstroms.

Natural Light: Light from the sun.

Necrosis: Death of plant tissue.

Node: Place on the stem where leaves, buds, or flowers are attached.

Orangery: Structure, usually glass, in which citrus trees are overwintered.

Organic Fertilizer: Fertilizer in which the essential elements are derived from once-living materials as a result of decomposition.

Organic Matter: Refuse derived from decaying material which was once part of living plants or animals.

Overwatering: Injury, or death, of plant roots due to lack of oxygen. Occurs as the result of the pores of the growing medium being filled with water, either due to poor drainage or too frequent application.

Parasitoid: A biological control in which the beneficial lays an egg on or in its victim. The larva that hatches feeds on its host, ultimately killing it.

Pathogen: Disease-causing organism, including viruses, fungi, bacteria, mycoplasma-like organisms, and viroids.

Peat-Lite Mixes: Growing media composed of sphagnum peat moss and vermiculite and/or perlite.

Peristylium: Court enclosed by columns.

pH: Index of the acidity or alkalinity of the growing medium. pH ranges from 1.0, which is extremely acid, to 14.0, which is alkaline; 7.0 is neutral.

Photoperiod: Length of the day or hours of illumination and its influence on plants.

Photosynthesis: Production of sugar in green plants from carbon dioxide and water in the presence of chlorophyll and light.

Phytochrome: Pigment of plants associated with certain red/far-red responses, including photoperiodism, seed germination, and onset of dormancy.

Phytotoxic: Toxic, or harmful, to plants.

Predator: A biological control that catches and consumes its pray.

Preventive Control: *See* Cultural Control.

Proportioner: Machine that mixes water and fertilizer from a concentrated stock solution to produce a final solution at the desired concentration.

Psychrometer: Instrument used to determine relative humidity by reading two thermometers, the bulb of one of which is kept moistened and ventilated.

Pycnidia: Asexual fruiting body in certain fungi, which produces spores in chains.

Relative Humidity: Actual humidity of the air compared to the potential humidity (vapor pressure) of the air at a given temperature.

Respiration: Oxidation of food, with the release of energy. Provides energy for processes of life, and releases carbon dioxide and water.

Restricted-Use Pesticides: All products that bear a Poison-Danger statement and a skull and crossbones on the label, except antimicrobial agents such as disinfectants, sanitizers, preservatives, and human and veterinary health products.

Rotational Planting: Flowering plants in an interior plantscape which are changed on a regular basis to maintain color year round.

Salinity meter: Instrument that measures conductivity of the soil solution. Used to determine soluble-salt levels in the growing medium.

Sclerotium: Vegetative, resting, food-storage body of certain fungi.

Senescence: Process of aging.

Short-Day Plant: Responds when the day length is less than a certain critical number of hours.

Shrinkage: The reduction in volume of a container medium; usually results in increased water retention and decreased aeration.

Sick Building Syndrome: Human health problems related to working or being in a building and generally associated with air pollution problems.

Sodium Lamp: High-energy lamp producing light by passing a discharge through sodium vapor.

Soil Structure: Arrangement of the soil particles.

Soil Texture: Type of particles present—sand, silt, clay.

Soluble Salts: Soluble minerals in the growing medium. Come from parent minerals of the medium, applied fertilizer, and irrigation water.

Spectral-Emission Discharge (S.E.D.): Wavelengths of radiation emitted by a light source, producing its characteristic color.

Spectrum: Band of colors produced when a beam of light is passed through a prism. White light, the sun, produces a band ranging from red to orange, yellow, green, blue, and violet.

Stomate: Pore in the leaf for the exchange of gases.

Subirrigation: Irrigation of plants from the bottom which depends upon the capacity of a growing medium to lift water against gravity by capillarity.

Supplemental Light: Use of artificial light together with natural light to raise light intensity to the desired level.

Top Dressing: Dry fertilizer is applied to the surface of the soil.

Trace Elements: *See* Micro Nutrients.

Translocation: Movement of materials throughout the plant.

Transpiration: Evaporation of water from the cells of plants and its subsequent diffusion into the air.

Turgor: Rigidity of cells, or whole plants, due to the pressure of water against the cell walls.

U.C. Mixes: Growing media composed of fine sand and sphagnum peat in varying ratios.

Wardian Case: Glass-enclosed growing case providing an ideal environment for plants requiring high humidity. Introduced by N. B. Ward in 1831.

Water Quality: A measure of all the dissolved and suspended substances in water.

Watts-per-Square-Meter Irradiance: Method of expressing light intensity at a specified distance from a particular illumination source.

Wettable Powder (WP): Pesticide or fungicide that forms a suspension of chemical particles when mixed with water.

Wetting Agent: Material added to the growing medium or irrigation water, to increase the ability of water to wet the particles.

Wilting Point: Amount of water in the growing medium below which plants will not recover from wilting unless additional water is added.

Worker Protection Standard: A safety measure designed to reduce pesticide-related illnesses and injuries to owners and managers of agricultural and horticultural businesses, their employees, pesticide applicators, and consultants.

WPS: *See* Worker Protection Standard.

APPENDIX A

Some Sources of Decorative Planters

ACURA FIBERGLASS CORP. P.O. Box 186, Palmetto, FL 34221.

AKRO-MILLS SPECIALTY PRODUCTS 1293 South Main Street, Akron, OH 44301.

ARCHITECTURAL BRASS CO. 996 Huff Road N.W., Atlanta, GA 30318.

ARCHITECTURAL POTTERY 15161 Van Buren, Midway City, CA 92655.

ARCHITECTURAL PRECAST, INC. P.O. Box 23110, Columbus, OH 43223.

ARCHITECTURAL SUPPLEMENTS 93 Triangle Street, Danbury, CT 06810.

BEACH POLYCERAMICS, INC. 156 West 71st Avenue, Vancouver, BC, Canada V5X 4S7.

BEMIS MANUFACTURING COMPANY, Decor Division, 300 Mill Street, P.O. Box 901, Sheboygan Falls, WI 53085-0901.

BONAR PLASTICS, INC. 1005 Atlantic Avenue, West Chicago, IL 60185.

THE BROOKFIELD COMPANY P.O. Box 80424, Chattanooga, TN 37414.

CAMPANIA, INC. 1320 North West End Boulevard, Route 309, Quakertown, PA 18951.

CANTERBURY INTERNATIONAL 5632 West Washington Boulevard, Los Angeles, CA 90016-1986.

CANYON POTTERY COMPANY, INC. 3621 Bandini Street, San Diego, CA 92110.

CHEM-TAINER INDUSTRIES, INC. 316 Neptune Avenue, North Babylon, NY 11704.

CLAYCRAFT PLANTER CO., INC. 807 Avenue of the Americas, New York, NY 10001.

COLUMBIA CASCADE CO. 1975 S.W. Fifth Avenue, Portland, OR 97201-5293.

COUNTRY CASUAL 17317 Germantown Road, Germantown, MD 20874-2999.

DARBRO PLANTERS 356 State Place, Escondido, CA 92025.

DAVIDSON-UPHOFF, INC. P.O. Box 184, Clarendon Hills, IL 60514.

DIAL INDUSTRIES 1538 Esperanza Street, Los Angeles, CA 90023.

DURA ART STONE 11010 Live Oak, Fontana, CA 92337.

DURACO PRODUCTS 1109 East Lake Street, Streamwood, IL 60107-4395.

DURAFORM 1435 South Santa Fe Avenue, Compton, CA 90221.

DUTCH PRODUCT & SUPPLY CO. 166 Lincoln Avenue, Yardley, PA 19067.

DYNAMIC DESIGN, INC. 19-40 Hazen Street, East Elmhurst, NY 11370.

EARTHSTONE 4139 East Bell Road, Phoenix, AZ 85032.

ENVIRONMENTAL FEATURES, INC. 21095 Halsted Road, Northville, MI 48167.

FIB-CON CORPORATION Box 3387, Silver Spring, MD 20901.

FIBERGLASS ENGINEERING CO. P.O. Box 117, Midland, VA 22728.

FIBERGLASS FABRICATORS 456 Montgomery Street, Orange, CA 92668.

FORMS AND SURFACES Box 5215, Santa Barbara, CA 93150.

FRANKLIN CHINA COMPANY, INC. 816 Nina Way, Warminster, PA 18974.

GAINEY CERAMICS, INC. 1200 Arrow Highway, LaVerne, CA 91750.

GLADDING MCBEAN POTTERY 1345 Toleman Creek Road, Ashland, OR 97520.

GLASSPEC CORPORATION 10344 S.W. 187th Street, Miami, FL 33157.

GOODWIN INTERNATIONAL 2915 D-101, Redhill Avenue, Costa Mesa, CA 92626.

GRUBB POTTERY MANUFACTURING CO. 2800 West Grand Avenue, Marshall, TX 75670.

HADDONSTONE USA LTD. 201 Heller Place, Bellmawr, NJ 08031.

HALL-WOOLFORD TANK COMPANY, INC. 5500 North Water Street, P.O. Box 2755, Philadelphia, PA 19120-0755.

HARRIS POTTERIES 800 North Wells Street, Chicago, IL 60610.

HINES III, INC. 3621 St. Augustine, Jacksonville, FL 32207.

HINES POTTERY 6747 Signat Drive, Houston, TX 77041.

HOSLEY BRASS 20530 Stoney Island Avenue, Lynwood, IL 60411.

INFINITY FIBERCORP 100 North Perry Lane, Tempe, AZ 85281.

INNISFREE, LTD. 6955 N.W. 36th Avenue, Miami, FL 33147.

INTERIOR GARDENS PRODUCTS 75 Darling Avenue, South Portland, ME 04106.

INTERNATIONAL TERRA COTTA, INC. 690 North Robertson Boulevard, Los Angeles, CA 90069-5088.

JACKSON'S POTTERY 6950 Lemmon Avenue, Dallas, TX 75229.

JA-LU DISTRIBUTING CO. 120 Candace Drive, Maitland, FL 32751.

JAZ PRODUCTS P.O. Box 3504, Thousand Oaks, CA 91359.

KI P.O. Box 8100, Green Bay, WI 54308-8100.

KOBA CORPORATION 60 Baekeland Avenue, Middlesex, NJ 08846.

KOLLER FIBERGLASS 9194 Davenport Street N.E., Blaine, MN 55434.

LANDSCAPE FORMS, INC. 431 Lawndale Avenue, Kalamazoo, MI 49001-9543.

LAWRENCE METAL PRODUCTS, INC. P.O. Box 400-M, Bay Shore, NY 11706.

THE LERIO CORPORATION P.O. Box 2084, Mobile, AL 36652.

MAGNALITE SYSTEMS, INC. 2900 Lockheed, Carson City, NV 89706.

MARSHALL POTTERY 1137 Conveyor, Suite 118, Dallas, TX 75247.

METAL WEAVE PRODUCTS CORP. 111 Cedar Street, New Rochelle, NY 10801.

MID-ATLANTIC POTTERY, INC. P.O. Box 246, La Plata, MD 20646.

MOLDED FIBERGLASS TRAY CO. East Erie Street, Linesville, PA 16424.

NEVINS INTERNATIONAL 9337B Katy Freeway, #320, Houston, TX 77024.

NEW ENGLAND POTTERY U.S. Route 1, Foxboro, MA 02035.

NICHOLS BROS. STONEWORKS 20209 Broadway, Snohomish, WA 98290.

NILAND COMPANY 7241 Stiles, El Paso, TX 79915.

NORCAL POTTERY PRODUCTS, INC. P.O. Box 1628, San Leandro, CA 94577.

NURSERY SUPPLIES, INC. 1415 Orchard Drive, Chambersburg, PA 17201.

OLCOTT INDUSTRIES 2051 Transit Road, Burt, NY 14028.

PALECEK P.O. Box 225, Station A, Richmond, CA 94808-0225.

PAM POTTERY SALES, INC. 8800 N.W. 15th Street, Miami, FL 33172.

PETER PEPPER PRODUCTS, INC. 17929 South Susana Road, Comptom, CA 90224.

PICOPOTAMUS, INC. 34 Futurity Gate, Unit 11-12, Concord, ON, Canada L4K 1S6.

THE POT COMPANY 2601 N.W. 112th Avenue, Miami, FL 33172-1804.

POT SPECIALISTS 1 West Street, Fall River, MA 02720.

POTTERY MANUFACTURING AND DISTRIBUTION, INC. 18881 South Hoover Street, Gardena, CA 90248.

PRIMESCAPE PRODUCTS COMPANY P.O. Box 710, Deerfield, IL 60015.

PYRO MEDIA 7911 10th Avenue South, Seattle, WA 98108-4404.

QUICK CRETE PRODUCTS CORP. P.O. Box 639, Norco, CA 91760.

RIVERSIDE PLASTICS, INC. P.O. Box 421, Flemingsburg, KY 41041.

ROBERT CHARLES PRODUCTS P.O. Box 868, West Caldwell, NJ 07707-0868.

ROBINSON CONSUMER PRODUCTS 205 Armstrong Road, Des Plaines, IL 60018.

ROBINSON IRON P.O. Box 1119, Alexander City, AL 35010.

ROTOCAST PLASTIC PRODUCTS, INC. 6546 N.W. 67th Street, Miami, FL 33147.

RPI DESIGNS 7079 Peck Road, Marlette, MI 48453.

R. P. PROFILES CORPORATION P.O. Box AF, Jackson, NJ 08527.

RUBBERMAID HORTICULTURAL PRODUCTS, INC. 3124 Valley Avenue, Winchester, VA 22601.

SCULPTURE DESIGN IMPORTS, INC. 416 South Robertson Boulevard, Los Angeles, CA 90048.

SITECRAFT, INC. 40-25 Crescent Street, Long Island City, NY 11101.

STONE YARD, INC. 9985 Huennekens Street, Suite B, San Diego, CA 92121.

SUMMIT FURNITURE, INC. 5 Harris Court, Building W, Monterey, CA 93940.

SUNSHINE PRODUCTS, INC. 307 North Figueroa Street, Wilmington, CA 90744.

SYRACUSE POTTERY, INC. 6551 Pottery Road, Warners, NY 13164.

TERRACAST 4700 Mitchell Street, North Las Vegas, NV 89031.

TERRA-FORM P.O. Box 1520, Wausau, WI 55402-1520.

TOPSIDERS 11325 Reed Hartmen Highway, #120, Cincinnati, OH 45241.

US POTTERY MANUFACTURING, INC. 15147 Colorado Avenue, Paramount, CA 90723.

ULTRUM P.O. Box 121, Fort Payne, AL 35967.

UNITED MARKETING, INC. P.O. Box 870, Pottsville, PA 17901-0870.

VALLEY VIEW SPECIALTIES CO. 13834 South Kostner, Crestwood, IL 60445.

VAST AMERICA, INC. 2651 Perth, Dallas, TX 75220.

VISTA PRODUCTS, INC. 28457 North Ballard Drive, Lake Forest, IL 60045.

WOODPECKER PRODUCTS 148 South G Street, Arcata, CA 95521.

ZANESVILLE STONEWARE COMPANY 1107 Muskingum Avenue, Zanesville, OH 43702-0605.

Self-Watering Planters

FENSMORE HYDROCULTURE PLANTSYSTEMS P.O. Box 2188, Apopka, FL 32704-2188.

GROSFILLEX Old West Penn Avenue, Robesonia, PA 19551.

JARDINIER PLANTER SYSTEMS, INC. 951 South Cypress, Suite M. LaHabra, CA 90631 (Jardinier).

KPA PRODUCTS 430 Evans Road, Milpitas, CA 95035 (Aqua Disk, Everlife Subirrigation, Mona Plant Systems).

PLANTER TECHNOLOGY 999 Independence Avenue, Mountain View, CA 94043-2302 (Natural Spring).

PRIMESCAPE PRODUCTS COMPANY P.O. Box 710, Deerfield, IL 60015 (Mona Plant Systems, Waterdisc).

TROPICAL ORNAMENTALS 5346 Woodland Drive, Delray Beach, FL 33484 (Everlife Subirrigation).

B

Some Leaf and Stem Diseases of Foliage Plants

AECHMEA FASCIATA

Anthracnose (*Colletotrichum* sp.)
See *Aspidistra*.

Bacterial Leaf Spot (*Erwinia carotovora* subsp. *carotovora*)
See *Aglaonema*.

Exserohilum Leaf Spot (*Exserohilum rostratum*)
See Palms, Bipolaris Leaf Spot.

Helminthosporium Leaf Spot (*Helminthosporium rostratum*)
This disease begins as tiny, water-soaked spots, yellowish in color. The spots enlarge, with the chlorotic areas becoming sunken and brown. Under optimum conditions, spots coalesce and entire leaves may collapse and hang limply from the plant. Death of infected leaves may occur.

Rhizoctonia Leaf Spot (*Rhizoctonia solani*)
Foliage plants have variable symptoms, depending upon the plant species and the environment. Usually recognized by the mass of reddish-brown mycelia (plant body) which resembles a spiderweb. On certain hosts, discrete lesions may develop. Fungus infests leaves, stems, and roots.

AESCHYNANTHUS PULCHER

Botrytis Blight (*Botrytis cinerea*)
    May develop on leaves near the pot rim or in contact with the growing medium. A water-soaked lesion forms which may enlarge to encompass a large portion of the leaf. Lesions turn necrotic and become black or brown as they age. Flowers may also be infected.

Corynespora Leaf Spot (*Corynespora cassicola*)
    Tiny, slightly brown lesions enlarge to 3–5 mm in diameter. Spots have a bright purple to red margin surrounded by a yellow halo about 1 mm wide.

Myrothecium Leaf Spot (*Myrothecium roridum*)
    See *Aphelandra*.

Phytophthora Stem Rot (*Phytophthora parasitica*)
    See *Aphelandra*.

Tobacco Mosaic (Tobacco Mosaic Virus)
    Pale yellowish streaks and patches develop on the upper surfaces of leaves which are distorted and stunted. Stems may branch excessively.

AGAVE

Anthracnose (*Glomerella cingulata*)
    Circular, depressed, dark-colored spots up to 1 in. in diameter bordered by a raised ring. Spots may run together and destroy the leaf.

Leaf Blight (*Botrytis cinerea*)
    See *Ficus*.

Leaf Spot (*Coniothyrium concentricum*)
    Zoned, light grayish-brown spots 1 in. or more in diameter. Concentric rings of the fruiting bodies of the fungus with light brown spores are conspicuous within the spots. Large portions of the leaf may be destroyed.

AGLAONEMA

Anthracnose (*Colletotrichum* sp.)
    See *Aspidistra*.

Bacterial Leaf Blight (*Xanthomonas dieffenbachiae*)
    See *Syngonium*.

Bacterial Leaf Spot (*Pseudomonas cichorii*)
    See *Dracaena*.

Bent Tip of 'Silver Queen'
    Frequently exhibited by *A. commutatum* 'Silver Queen', the tip of a new leaf

may be bent back flat against the leaf. As the leaf unfurls, the area at the base of the bend may tear and the leaf tip fall off. The bent portion varies from 0.1 to 4 cm. The cause is unknown. Low light, moist growing medium, and moderate fertilizer rates reduce incidence.

Cripple Leaf

Leaves of *A. commutatum* 'Fransher' are distorted. New leaves small and distorted, sometimes slightly curved or hooked, with edges rolling toward the center. Copper deficiency is the cause. Treatment with copper sequestrene will correct the problem.

Dasheen Mosaic Virus

This pathogen appears to infect only members of the family *Araceae*. A mosaic pattern on the foliage is the most common symptom. Infected plants are stunted and disfigured, but rarely killed.

Erwinia Blight and Stem Rot *(Erwinia chrysanthemi, E. carotovora)*

Symptoms of this disease following internal invasion by the pathogen include foliar yellowing of new leaves, often with accompanying wilt, followed by a mushy, foul-smelling stem rot. Foliar infections cause a rapid, mushy leaf collapse. Spots may be surrounded by a yellow halo.

Destroy diseased plants and keep foliage dry. This is the single, most important bacterial phytopathogen of tropical foliage plants.

Fusarium Stem Rot *(Fusarium solani)*

Leaf spots are tan and dry with concentric rings of light and dark tissue, and may be up to 3 cm wide. On petioles, lesions are elongated and sunken with the characteristic purplish margins associated with *Fusarium* stem lesions.

Myrothecium Leaf Spot *(Myrothecium roridum)*

Leaf spots are usually found at a wound site and may be up to 1 in. in diameter. Spots are tan to brown and may have a yellow border. The lower leaf surface shows the black and white fruiting bodies of the fungus in concentric rings near the outer edge of the spot.

Root and Stem Rot *(Rhizoctonia, Phythium, and Phytophthora, Fusarium)*

See text, page 213.

ANTHURIUM

Anthracnose *(Gloeosporium minutum)*

Found occasionally in greenhouses. More-or-less circular spots along veins run together to form necrotic areas 1 or more inches in diameter surrounded by a yellow border. Diseased tissue dries and falls out.

Bacterial Blight *(Xanthomonas dieffenbachiae)*

See *Syngonium*.

Bacterial Leaf Spot *(Pseudomonas cichorii, Xanthomonas dieffenbachiae)*

See *Dracaena*.

Colletotrichum Leaf Spot (*Colletotrichum* sp.)
See *Aspidistra.*

Dasheen Mosaic Virus
See *Aglaonema.*

Leaf Spot (*Cephaleuros parasiticus*)
Caused by an algae. Dark gray spots ½ in. in diameter develop along the main veins. Veins become corky.

Rhizoctonia Aerial Blight (*Rhizoctonia solani*)
See *Fittonia.*

Soft Rot (*Erwinia carotovora* subsp. *carotovora*)
See *Algaonema.*

APHELANDRA

Aphelandra Stem Gall (*Nectriella pironii*)
Round, corky galls up to 2 in. in diameter form where tissues have been wounded. Destroy infected plants.

Bacterial Leaf Spot (*Xanthomonas campestris*)
See *Dracaena.*

Botrytis Blight (*Botrytis cinerea*)
Large dark green to gray areas primarily along the leaf edges. The dusty gray-tan spores are easily seen with a 10× hand lens. Infected leaves generally collapse.

Corynespora Leaf Spot (*Corynespora cassiicola*)
Leaf spots which start on edges, tips, and sometimes centers, are dark brown to black and may appear wet.

Mosaic (Cucumber mosaic virus)
Symptoms appear as bright yellow ringspot or mosaic pattern on infected leaves. Plant growth is slowed.

Myrothecium Leaf Spot (*Myrothecium roridum*)
Leaf spots are similar to *C. cassiicola* from the upper surface. Leaf undersides frequently show the fungal fruiting bodies in concentric rings within dead spots. Fruiting bodies are irregularly shaped black structures with a white fringe.

Phytophthora Stem Rot (*Phytophthora parasitica*)
Usually starts at the soil line and causes a blistering of the stem surface. The lesions are black and slightly mushy and may extend from the base of the stem up into the petioles of the lower leaves. Complete collapse of the plant may occur.

Southern Blight (*Sclerotium rolfsii*)
See *Brassaia.*

ARAUCARIA

Blight (*Cryptospora longispora*)
    The disease infects the lower branches first, and spreads upward. As the entire branch becomes infected, the tip becomes bent. Limbs die and break off from the tips. Plants will die if left untreated. Prune and burn diseased branches.

Colletotrichum Needle Blight (*Colletotrichum derridis*)
    Needle necrosis occurs primarily on newly developing needles and appears as tan to brown areas frequently accompanied by black fruiting bodies of the fungus.

ASPARAGUS

Blight (*Ascochyta asparagina*)
    Branches become dry and drop off. When severe, plants may be killed down to the crown.

Crown Gall (*Erwinia tumefaciens*)
    Pale green galls up to 2 in. in diameter are formed at the base of the stem. Galls are thick, fleshy, and irregular clumps developed from secondary sprouts.
    Remove and destroy the galls. Do not propagate from diseased plants.

ASPIDISTRA

Leaf Spots
    (*Colletotrichum omnivorum*)   Large white spots with brown margins on leaf blades and petioles is a typical symptom of this disease.
    (*Ascochyta aspidistrae*)   Large, irregular, pale spots on the leaves typify the presence of this pathogen.

ASPLENIUM

Bacterial Leaf Spot (*Pseudomonas cichorii, P. gladioli*)
    Small, water-soaked, translucent spots form all over the leaves. Lesions enlarge rapidly to $\frac{1}{8}$ in. in diameter and become reddish brown with a purple halo. Under warm, wet conditions, lesions may collapse, coalesce, and spread along veins to encompass a large portion of the fronds. Spots may be vein delimited and do not cross the central vein.

Foliar Nematode (*Aphelenchoides fragariae*)
    Necrotic areas, the result of nematode feeding, form between the veins. Usually appear at the midveins and expand to encompass entire leaves.

Rhizoctonia Aerial Blight (*Rhizoctonia solani* AG1)
    See *Fittonia*.

BEGONIA

Bacterial Blight *(Xanthomonas begoniae)*
The initial symptom is blister-like lesions on leaves. As they mature, the lesions become roughly circular and necrotic, and coalesce. Water-soaked margins may develop. Premature abscission of leaves is common.

Botrytis Blight *(Botrytis cinerea)*
See *Aphelandra.*

Corynespora Leaf Spot *(Corynespora cassiicola)*
See *Zebrina.*

Myrothecium Leaf Spot *(Myrothecium roridum)*
See *Aphelandra.*

Powdery Mildew *(Erysiphe cichoracearum)*
See *Kalanchoe.*

Rhizoctonia Stem Rot *(Rhizoctonia solani AG4)*
See *Fittonia.*

Soft Rot *(Erwinia carotovora* subsp. *carotovora)*
See *Aglaonema.*

BRASSAIA (SCHEFFLERA)

Alternaria Blight *(Alternaria actinophylla;* Figure B.1)
Infected leaves show brownish-black lesions that vary from small and circular to those that are varied in shape and cover most, if not all of the leaf. Usually, a

Figure B.1   *Alternaria* blight of *Brassaia.* (Courtesy of Foliage Education and Research Foundation, Inc., Apopka, FL).

yellow halo surrounds the diseased area. Infected leaves usually drop. In severe cases, petioles and stems may become infected. Black spores of the fungus are usually evident.

Bacterial Leaf Spot (*Pseudomonas cichorii, Erwinia chrysanthemi*)
    See *Schefflera*.

Cercospora Leaf Spot (*Cercospora* sp.)
    Pinpoint swellings produced on the undersurface of the leaf are the characteristic symptoms of this disease. Edema-like symptoms resemble those of improper water relations. A chlorotic border may be produced around the red to brown lesions, the leaf later turning yellow and abscising.

Colletotrichum Leaf Spot (*Colletotrichum* sp.)
    See *Aspidistra*.

Phytophthora Leaf Spot (*Phytophthora parasitica* var. *nicotianae*)
    Translucent spots enlarge to form large, irregular, brown, water-soaked lesions, often accompanied by some leaf abscission.

Rhizoctonia Aerial Blight (*Rhizoctonia solani*)
    See *Fittonia*.

Southern Blight (*Sclerotium rolfsii*)
    This disease appears as a brown stem rot completely encircling the stem at the soil line. The course mycelium of the fungus, intersperced with small, tan, spherical sclerotia is visible within the necrotic tissue. The plant may collapse at the soil line.

Xanthomonas Leaf Spot (*Xanthomonas campestris* pv. *hederae*)
    See *Hedera*.

CACTUS

Anthracnose (*Mycosphaerella opuntiae*; Figure B.2)
    On *Opuntia*, this disease causes a light brown rot. Light pink pustules containing spores develop on the surface of infected stems.

Corky scab
    A physiological disease. Irregular rusty, corky scabs develop and may destroy entire shoots. *Opuntia* is particularly susceptible. Increasing light and decreasing humidity will control.

Drechslera Stem Rot (*Drechslera cactivora*)
    Older plants become rotted where spines have broken or the stem punctured. Plants usually become blackened and may be mushy or dry. Affected portions usually collapse.

Dry Rot (*Phyllosticta concava*)
    Begins as small, black, circular spots which slowly increase to 1 to 2 in. in

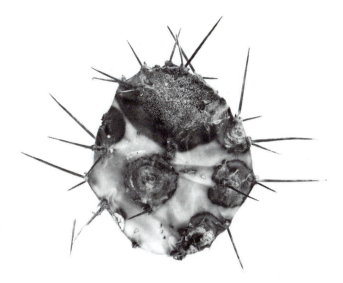

Figure B.2   Anthracnose of *Opuntia*. [From P. P. Pirone, *Diseases and Pests of Ornamental Plants*, 5th ed., John Wiley & Sons, Inc., New York, 1978.]

diameter. The diseased area becomes sunken with conspicuous black fruiting bodies dotting the surface.

Erwinia Stem Rot (*Erwinia* sp.)
A mushy, foul-smelling rot.

Soft Rot, Gray Mold (*Botrytis cinerea*)
Segments are discolored and the upper surface gradually rots and collapses. Black sclerotia may be conspicuous.

Virus (Cactus virus X)
Infected plants are stunted. Destroy diseased plants.

CALATHEA

Alternaria Leaf Spot (*Alternaria alternata*)
A disease characterized by small (⅛ in.), reddish brown, circular lesions with a yellow halo. Lesions are smaller than those caused by *Drechslera setariae*.

Drechslera Leaf Spot (*Drechslera setariae*)
Lesions are ⅛ to ½ in. wide, irregularly shaped, tan colored, with yellow border.

Mosaic (Cucumber mosaic virus)
See *Aphelandra*.

CISSUS

Anthracnose (*Colletotrichum* sp.)
See *Aspidistra*.

Botrytis Blight (*Botrytis cinerea*)
See *Aphelandra*.

Powdery Mildew (*Oidium* sp.)
See *Kalanchoe*.

Pseudomonas Leaf Spot (*Pseudomonas chichorii*)
See *Schefflera*.

Rhizoctonia Aerial Blight (*Rhizoctonia solani*)
See *Fittonia*.

Southern Blight (*Sclerotium rolfsii*)
See *Brassaia*.

CODIAEUM

Anthracnose (*Glomerella cingulata*)
Large, yellowish-gray spots usually appear on the upper side of the leaves when this pathogen is present. As the disease progresses, spots turn whitish and dry out. Salmon-colored spore pustules may be evident on the spots.

Bacterial Leaf Spot (*Xanthomonas campestris* pv. *poinsettiae*)
See *Syngonium*.

Crown Gall (*Agrobacterium tumefaciens*)
Slightly swollen areas on stems, leaf veins, and roots enlarge and become corky. In severe infections, galls merge to create a very distorted stem or root mass. Destroy infected plants.

Stem Gall and Canker (*Nectriella pironii*)
Large stem galls are near-spherical, corky, rough areas up to 48 mm in diameter. Galls can also form on petioles and midveins of leaves.

CORDYLINE

Anthracnose (*Glomerella cingulata*)
See *Codiaeum*.

Cercospora Leaf Spot
See *Brassaia*.

Erwinia Stem and Root Rot (*Erwinia chrysanthemi*, *E. carotovora* subsp. *carotovora*)
See *Aglaonema*.

Fusarium Leaf Spot (*Fusarium moniliforme*)
See *Dracaena*.

Leaf Spot (*Phytophthora parasitica*)
Spots on lower leaves exposed to splashing of water are irregular, necrotic, and zonate and may have water-soaked, advancing margin. This disease also infects the unfurled apical leaf if it is moist.

Phyllosticta Leaf Spot (*Phyllosticta draconis*)
Occurs primarily on older leaves and begins as a yellow area, which later turns brown with a purple margin, and frequently a yellow halo. Spots may coalesce and infected leaves die. The black fruiting bodies of the fungus form concentric rings within the lesions.

Southern Blight (*Sclerotium rolfsii*)
See *Brassaia*.

CRASSULA

Cercospora Leaf Spot
See *Brassaia*.

DIEFFENBACHIA

Anthracnose
See *Codiaeum*.

Bacterial Leaf Spot (*Xanthomonas dieffenbachiae*)
Spots are small at first, yellow or yellowish-orange with translucent centers. Bacterial ooze may appear on the lower surface of the spots which forms a waxy, silvery-white layer as it dries. Spots enlarge to ½ in. and may run together in moist conditions. In dry conditions, spots do not enlarge but turn reddish brown, making leaves appear speckled. If severe, yellowing, wilting, and death of infected parts will occur.

Brown Leaf Spot (*Leptosphaeria* sp.; Figure B.3)
Spots range from yellowish to grayish brown and may be found on the leaf blade, midveins, petiole, and flower spathe. Mature leaves are more readily infected than young leaves, and the lower leaf surface more than the upper. Spots may be extremely numerous, and range in size from pinpoint to 25 mm. Black fruiting structures are prominent in the centers of older lesions. The leaves gradually become chlorotic and die.

Dasheen Mosaic Virus
See *Aglaonema*.

Erwinia Blight and Stem Rot (*Erwinia* sp.)
Brownish, water-soaked, sunken, soft areas develop on the stem, with irregu-

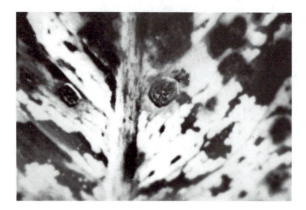

Figure B.3   Brown leaf spot of *Dieffenbachia*. (Courtesy of Foliage Education and Research Foundation, Inc., Apopka, FL).

lar, brownish, soft spots apparent on the leaves. A rotten, fishy odor may be apparent and a bacterial slime may be present.

Fusarium Stem Rot *(Fusarium oxysporum, F. solani)*
A soft, mushy rot at the base of the stem. Rotted area often has a purplish to reddish margin. Under conditions of high moisture, leaves may also be infected resulting in tan, papery spots up to 3 cm in diameter with concentric rings.

Glomerella Leaf Spot *(Glomerella cingulata)*
Leaf spots are initially water-soaked and may have a bright yellow halo. Fruiting bodies appear in concentric rings of black specks within the spot.

Leaf Spot *(Cephalosporium dieffenbachiae)*
Spots appear as small, reddish-brown lesions on the younger leaves and may be ¼ in. in diameter or more, with a dark border when the leaves are unrolled. Lesions run together, and yellowing and death of the leaf may occur.

Myrothecium Leaf Spot *(Myrothecium roridum)*
See *Aphelandra*.

Phytophthora Leaf Spot *(Phytophthora sp.; Figure B.4)*
Early symptoms are similar to bacterial infection, with mushy, shiny, water-soaked spots. In moist situations, the spot enlarges to encompass most of the leaf. Upon drying, lesions become a light, reddish tan with a yellow halo.

Pseudomonas Leaf Spot *(Pseudomonas cichorii)*
See *Schefflera*.

Southern Blight *(Sclerotium rolfsii)*
See *Brassaia*.

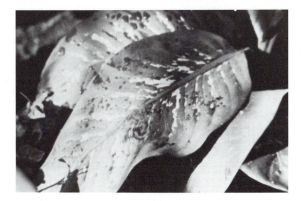

Figure B.4 *Phytophthora* leaf spot of *Dieffenbachia*. (Courtesy of Foliage Education and Research Foundation, Inc., Apopka, FL).

DRACAENA

Bacterial Leaf Spot *(Erwinia carotovora* subsp. *carotovora, E. chrysanthemi)*
    See *Aglaonema*.

Bacterial Leaf Spot of D. sanderana *(Pseudomonas* sp.)
    Symptoms appear as circular to irregular water-soaked spots occurring anywhere on the leaf blade. A reddish-brown margin usually forms around the lesions, and a diffuse chlorotic pattern develops. Spots often enlarge, turning papery and dry.

Botrytis Blight *(Botrytis cinerea)*
    See *Aphelandra*.

Colletotrichum Leaf Spot *(Colletotrichum* sp.)
    See *Aspidistra*.

Fusarium Leaf Spot *(Fusarium moniliforme;* Figure B.5)
    Infection occurs only in the growing point, and may rot the apex. Leaf spots appear as circular or slightly raised, reddish-brown lesions surrounded by a yellow halo. Cream-colored spores of the fungus may be seen within the infected areas. *D. marginata* is a common host.

Leaf Spot *(Glomerella cincta)*
    Dark discoloration originates at the leaf tips and progresses toward the base.

Leaf Spot *(Phyllosticta maculicola;* Figure B.6)
    Irregular, small, brown spots with yellowish margins develop on the leaves.

Phyllosticta Leaf Spot *(Phyllosticta dracaenae)*
    A circular to irregular spot, 1 to 5 mm in diameter, is produced on the upper

Figure B.5　*Fusarium* leaf spot of *Dracaena reflexa*. (Courtesy of Foliage Education and Research Foundation, Inc., Apopka, FL.)

or lower surfaces of older leaves. Lesions commonly have brown centers with purple borders and yellow halos. When severe, foliar necrosis may occur.

Phytophthora Leaf Spot *(Phytophthora parasitica)*
　　See *Cordyline*.

Soft Rot *(Erwinia carotovora* subsp. *carotovora)*
　　See *Dieffenbachia*.

Figure B.6　*Phyllosticta* leaf spot of *Cordyline*. (Courtesy of Foliage Education and Research Foundation, Inc., Apopka, FL).

Southern Blight *(Sclerotium rolfsii)*
See *Brassaia*.

Stem Rot *(Aspergillus niger)*
See *Sansevieria*.

Tip Blight *(Physalospora dracaenae)*
The disease begins on the lower leaves as areas of shrunken, straw-colored tissue. Center leaves die only at the tips.

EPIPREMNUM

Bacterial Leaf Spot *(Erwinia sp., Pseudomonas cichorii)*
Lesions initially water-soaked, rapidly turning dark brown to black. Spots often have concentric rings of light and dark tissue and are sometimes surrounded by a yellow halo. Spots roughly circular, up to 3 cm in diameter.

Rapid Decay of Pothos *(Erwinia carotovora)*
The pathogen invades leaves and petioles of potted plants, and stems, roots, and leaves of cuttings. The infected area appears as a discrete water-soaked, grayish-green area at first, and rapidly enlarges, becoming mushy and brown to black. Complete collapse of infected tissue ultimately occurs. In a dry environment, leaf lesions will turn brownish black, often with a yellow margin.

Southern Blight *(Sclerotium rolfsii)*
See *Brassaia*.

FATSIA

Alternaria Leaf Spot *(Alternaria panax)*
See *Brassaia*.

Bacterial Leaf Spot *(Pseudomonas cichorii, Xanthomonas campestris* pv. *hederae)*
See *Schefflera*.

Rhizoctonia Aerial Blight *(Rhizoctonia solani)*
See *Fittonia*.

Southern Blight *(Sclerotium rolfsii)*
See *Brassaia*.

FICUS

Anthracnose *(Glomerella cingulata;* Figure B.7)
The disease may appear as a tip-burn, with the ends of leaves turning yellowish at first, then tan, and finally dark brown. The entire margin may be affected with the disease working inward until the entire leaf is destroyed. Rose-colored pustules,

Figure B.7   Anthracnose of *Ficus*. [From P. P. Pirone, *Diseases and Pests of Ornamental Plants*, 5th ed., John Wiley & Sons, Inc., New York, 1978.]

usually along the veins, may be present. Black ascocarps will also develop on the leaves.

Cercospora Leaf Spot
    See *Brassaia*.

Colletotrichum Leaf Spot (*Colletotrichum gloeosporioides*)
    See *Aspidistra*.

Corynespora Leaf Spot (*Corynespora cassiicola*)
    Small to large, reddish spots on the youngest mature leaves. Leaf abscission is common in severe infections.

Gray Mold (*Botrytis cinerea*)
Brown leaf spots with concentric rings develop on the leaves. At the growing tips, necrotic lesions may develop between the leaf sheath and the newly emerging leaf.

Leaf Spot (*Leptostromella elastica*)
First symptoms may appear as spots or streaks, spreading until the entire leaf is involved. Black lines outline spots on which colorless spores are produced by small, black pycnidia.

Mosaic (Fig mosaic virus)
In *F. nitida* leaves are reduced in size and develop mosaic. Destroy infected plants.

Phomopsis Twig Blight (*Phomopsis* sp.)
A problem of *F. benjamina*, symptoms are similar to those which occur when an improperly acclimatized plant is brought indoors. Shriveled, sunken bark is evident on infected branches following leaf loss. Infected plants may die. Phomopsis is a weak pathogen which causes injury when the host plant is stressed.

Southern Blight (*Sclerotium rolfsii*)
See *Brassaia*.

Xanthomonas Leaf Spot (*Xanthomonas campestris*)
See *Syngonium*.

FITTONIA

Mottle (Biden's mottle virus)
Symptoms include leaf distortion, usually more severe on one side of the leaf than the other. Plants are stunted.

Rhizoctonia Aerial Blight (*Rhizoctonia solani*)
Occurs at high temperatures and humidities. Brown, irregularly shaped lesions form all over the plant. The disease spreads rapidly and the reddish-brown, web-like mycelium can cover the entire plant.

Southern Blight (*Sclerotium rolfsii*)
See *Brassaia*.

Xanthomonas Leaf Spot (*Xanthomonas campestris*)
See *Syngonium*.

GYNURA

Sclerotinia Blight (*Sclerotinia sclerotiorum*)
Symptoms first appear on the leaves as a soft, brown decay often covered with white fungal mycelium. As the disease progresses, the stems become hollow and the plant may collapse. Black sclerotia may be found within the stem.

Figure B.8   Bacterial leaf spot of *Hedera*. [From P. P. Pirone, *Diseases and Pests of Ornamental Plants*, 5th ed., John Wiley & Sons, Inc., New York, 1978.]

HEDERA

Bacterial Leaf Spot and Stem Canker (*Xanthomonas hederae;* Figure B.8)
Leaf spots are light green and water-soaked at first, turning brown or black with reddish margins. Petioles become black and shriveled. On stems, a black decay occurs downward from the tip, and cankers may girdle the stem.

Botrytis Blight (*Botrytis cinerea*)
See *Aphelandra.*

Colletotrichum Leaf Spot (*Colletotrichum trichellum*)
Similar to *Xanthomonas.* Black lesions with black fruiting bodies in the center are typical symptoms.

Phyllosticta Leaf Spot (*Phyllosticta concentrica*)
See *Dracaena.*

Phytophthora Leaf Spot (*Phytophthora palmivora*)
Plants exhibit poor growth and color. Basal leaves turn brown and curl downward. Leaf spots are large, gray to black, and water-soaked. Root rot may also occur.

Southern Blight (*Sclerotium rolfsii*)
See *Brassaia.*

HOYA

Botrytis Blight (*Botrytis cinerea*)
See *Aphelandra.*

Cercospora Leaf Spot (*Cercospora* sp.)
See *Brassaia.*

Rhizoctonia Aerial Blight *(Rhizoctonia solani)*
   See *Fittonia*.

KALANCHOE

Cercospora Leaf Spot (*Cercospora* sp.)
   See *Brassaia*.

Myrothecium Leaf Spot *(Myrothecium roridum)*
   See *Aphelandra*.

Powdery Mildew *(Sphaerotheca humuli)*
   A grayish-white mealy growth develops on the leaves. Infected parts dry out and the leaves are killed. Many other species are susceptible to powdery mildew.

Soft Rot *(Erwinia carotovora* subsp. *carotovora)*
   See *Aglaonema*.

Stemphyllum Leaf Spot *(Stemphyllum bolicki)*
   Lesions on both leaf surfaces are dark brown to black and up to 3 mm in diameter. They appear raised and wart-like. Severe infections may result in chlorotic leaves and leaf abscission.

MARANTA

Biopolaris Leaf Spot *(Bipolaris setariae)*
   See *Palms*.

Helminthosporium Leaf Spot *(Drechslera setariae)*
   Small lesions, $\frac{1}{16}$ in. wide or less, first appear as tiny water-soaked areas which turn chlorotic and finally necrotic. Affected leaves have a speckled appearance. In severe cases, lesions coalesce to form irregularly shaped areas up to $\frac{1}{2}$ in. wide which are tan with a chlorotic halo.

Leaf Spot *(Glomerella cincta)*
   See *Dracaena*.

Mosaic Leaf Spot (Cucumber mosaic virus)
   See *Aphelandra*.

NEPHROLEPIS

Anthracnose *(Glomerella nephrolepis)*
   Soft growing tips of fronds turn brown and dry. Keep foliage dry; destroy diseased leaves.

Botrytis Blight *(Botrytis cinerea)*
   See *Aphelandra*.

Rhizoctonia Aerial Blight (*Rhizoctonia solani*)
　　See *Fittonia*.

PALMS

Bacterial Blight (*Pseudomonas alboprecipitans*)
　　Caryota mitis, fishtail palm, is particularly susceptible. Conspicuous symptoms are translucent water-soaked areas often surrounding brown lesions 1 to 2 to more than 50 mm in length, usually parallel to the veins. The interior of the lesions may become necrotic, with mature lesions often having a chlorotic margin. Leaves that are unfolding and immature leaves are most severely infected, although all leaves are susceptible. Removing infected foliage and keeping the leaves dry will control the disease.

Bipolaris Leaf Spot (*Bipolaris setariae*)
　　Pinpoint, water-soaked, grayish lesions scattered over the surface of the leaf enlarge to $\frac{1}{16}$ to $\frac{1}{8}$ in. within 2 to 3 weeks and may turn tan or reddish brown or black as they mature. A chlorotic halo may be present. Lesions often coalesce, with the pinnae becoming blackened and shredded. There is no chemical control. Growing unstressed plants is important.

Cercospora Leaf Spot
　　See *Brassaia*.

Cylindrocladium Leaf Spot (*Cylindrocladium* sp.)
　　Elongated, dark brown to black spots with a bright yellow halo. Lesions pinpoint to 1 cm in diameter.

Drechslera Leaf Spot (*Drechslera setariae*)
　　Spots appear on young, immature leaves as pinpoint, chlorotic lesions which enlarge and become elliptical, tan, or black with a halo. Lesions may coalesce and form large areas of necrotic tissue.

Exserohilum Leaf Spot (*Exserohilum rostratum*)
　　Similar to *Bipolaris*.

False Smut, Leaf Scab (*Graphiola phoenicis*)
　　This parasite causes a yellow spotting of the leaves and the formation of numerous black scabs, the fruiting bodies, the outer parts of which are dark, hard, and horny. Within, powdery yellow or light brown spore masses are produced. Infected leaves soon die.

Gliocladium Disease (*Gliocladium* sp.)
　　Typical symptom is the premature death of the oldest fronds. Necrotic streaks form from the base of the rachis outward, with pinnae often turning yellowish brown on one side of the rachis. Tissue under the necrotic areas is streaked with various shades of brown. The pinkish-brown sporulation of the fungus is readily visible.
　　Hosts include *Chamaedorea seifritzii*, *C. elegans*, *C. erumpens*, *Chrysalidocarpus*

lutescens, Archontophoenix cunninghamiana, Howea forsterana, Washingtonia filifera, W. robusta, Phoenix canariensis, and *Cocos plumosa.*

Leaf Spot *(Exosporium palmivorum)*

Spots are small, round, yellowish, and transparent, often running together to form large, irregular gray-brown blotches which may kill the leaf.

Phaeotrichoconis Leaf Spot *(Phaeotrichoconis crotalariae)*

Similar to *Bipolaris.*

PEPEROMIA

Colletotrichum Leaf Spot *(Collectotrichum* sp.)

See *Aspidistra.*

Crown Rot *(Phytophthora nicotianae)*

This disease appears as a crown rot, and is especially common on *P. obtusifolia variegata.*

Edema *(Cercospora* sp.; Figure B.9)

Intumescences on the lower surface of the leaf seldom exceed 2 mm in diameter. The slightly raised, subcircular, smooth or scablike swellings are dark green, with a necrotic tip up to 1 mm in height and prominent to the touch. Chlorotic spotting of the upper leaf may occur in severe infections. Older leaves are most susceptible.

Mosaic (Cucumber mosaic virus)

See *Aphelandra.*

Myrothecium Leaf Spot *(Myrothecium roridum)*

See *Aphelandra.*

Phyllosticta Leaf Spot *(Phyllosticta* sp.)

See *Dracaena.*

Figure B.9 *Cercospora* leaf spot of *Peperomia.* (Courtesy of Foliage Education and Research Foundation, Inc., Apopka, FL).

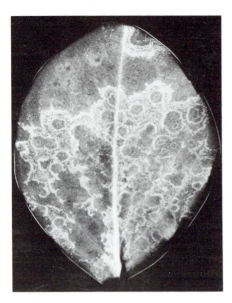

Figure B.10   Ringspot virus of *Peperomia*. [From P. P. Pirone, *Diseases and Pests of Ornamental Plants*, 5th ed., John Wiley & Sons, Inc., New York, 1978.]

Ringspot Virus (Figure B.10)

Affected plants are stunted and the leaves distorted. Chlorotic or necrotic rings appear in the leaves.

Southern Blight *(Sclerotium rolfsii)*

See *Brassaia*.

PHILODENDRON

Bacterial Leaf Spot *(Pseudomonas cichorii)*

Similar to *Erwinia* leaf spot, but the lesions rarely become mushy and do not appear water-soaked.

Red Edge *(Xanthomonas dieffenbachiae;* Figure B.11)

Xanthomonas is a common pathogen of *P. scandens* 'oxycardium'. Small water-soaked spots appear within the blade first, gradually turning yellow and enlarging to form irregular, elongated blotches. With age, the center of the lesion may turn brown. The most common symptom is yellowing of the leaf margin, with the tip infected first. The leaf margin becomes necrotic and turns reddish brown. Affected leaves will abscise.

Botrytis Blight *(Botrytis cinerea)*

See *Aphelandra*.

Cercospora Leaf Spot *(Cerospora* sp.)

Small yellow lesions which may reach 25 mm in diameter and develop sunken necrotic centers and chlorotic margins when mature.

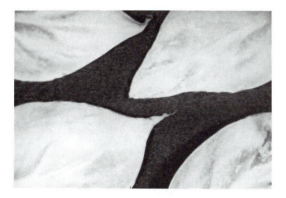

Figure B.11 *Xanthomonas* of *Philodendron*. (Courtesy of Foliage Education and Research Foundation, Inc., Apopka, FL).

Colletotrichum Leaf Spot (*Colletotrichum* sp.)
See *Aspidistra*.

Dasheen Mosaic Virus (Figure B.12)
In *Philodendron scandens* subsp. 'oxycardium,' the characteristic symptom is the formation of chlorotic bands in the leaves parallel to, but not limited by, veins. Leaf shape may be distorted; size is usually reduced. Mosaic symptoms are conspicuous on new leaves.

Figure B.12 Dasheen mosaic virus of *Philodendron scandens* 'oxycardium'. [From F. W. Zettler, G. C. Wisler, and J. J. McRitchie, "Dasheen Mosaic Virus of *Philodendron scandens* subsp. *oxycardium*," *Foliage Digest* 1(5): 15, 1978.]

Figure B.13 *Dactylaria* leaf spot of *Philodendron*. (Courtesy of Foliage Education and Research Foundation, Inc., Apopka, FL).

Dactylaria Leaf Spot (*Dactylaria humicola*; Figure B.13)

Young leaves are infected and show small, pinpoint, water-soaked spots on the undersurface. Spots may enlarge but are rarely greater than 2 mm in diameter. As infected leaves mature, lesions turn yellowish green to yellow, often with a brown, water-soaked center. The infected areas collapse, giving the lesions a concave appearance similar to thrips injury.

Erwinia Blight (Figure B.14)

See *Aglaonema*.

Myrothecium Leaf Spot (*Myrothecium roridum*)

See *Aphelandra*.

Physiological Leaf Spot

A leaf spot related to stomatal exudation and temperature is primarily a greenhouse problem. *P. domesticum* is very susceptible, particularly the older leaves. Exuding spots less than 2 mm in diameter with necrotic centers are the typical symptom. Sooty mold may grow on the exudate. High temperatures (75 to 80°F) favor this disorder.

Phytophthora Leaf Spot (Figure B. 15)

See *Dieffenbachia*.

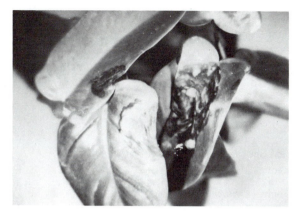

Figure B.14 *Erwinia* of *Philodendron*. (Courtesy of Foliage Education and Research Foundation, Inc., Apopka, FL).

Phytophthora Leaf Spot (*Phytophthora parasitica* var. *nicotianae*)
Irregularly shaped lesions are dark brown, water-soaked, and ½ to 1 in. wide.

Rhizoctonia Aerial Blight (*Rhizoctonia solani*)
See *Fittonia*.

Soft Rot (*Erwinia carotovora* subsp. *carotovora*)
See *Aglaonema*.

Southern Blight (*Sclerotium rolfsii*)
See *Brassaia*.

Figure B.15 *Phytophthora* leaf spot of *Philodendron*. (Courtesy of Foliage Education and Research Foundation, Inc., Apopka, FL).

PILEA

Bacterial Leaf Spot *(Xanthomonas campestris)*
  See *Schefflera*.

Cercospora Leaf Spot
  See *Brassaia*.

Colletotrichum Blight *(Colletotrichum capisci)*
  See *Aspidistra*.

Myrothecium Leaf Spot *(Myrothecium roridum)*
  See *Aphelandra*.

Phytophthora Blight *(Phytophthora parasitica)*
  See *Dieffenbachia*.

Rhizoctonia Aerial Blight *(Rhizoctonia solanii)*
  See *Fittonia*.

Southern Blight *(Sclerotium rolfsii)*
  See *Brassaia*.

PITTOSPORUM

Alternaria Leaf Spot *(Alternaria tenuissima)*
  Irregular, chlorotic spots are scattered over the leaf surface. Spots are usually round with brown or tan centers and may be surrounded by a yellow halo. Immature leaves may be crinkled.

Angular Leaf Spot *(Cercospora pittospori)*
  Light yellow to pale green and tan angular spots on upper leaf surfaces have irregular margins. The spot pattern is similar on the underside of the leaf.

Rough-Bark Disease
  Caused by a virus, the usual symptom is roughened bark. In severe cases, stems become girdled and swollen and the twig or branch above that point may die. The plant may be stunted with new leaves on infected branches distorted and having chlorotic areas of indefinite shape. Destroy infected plants.

POLYSCIAS (SCHEFFLERA)

Alternaria Leaf Spot *(Alternaria panax)*
  See *Brassaia*.

Bacterial Leaf Spot *(Pseudomonas cichorii, Xanthomonas campestris pv. hederae)*
  See *Dracaena*.

Southern Blight *(Sclerotium rolfsii)*
  See *Brassaia*.

RHOEO

Mosaic (Tobacco mosaic virus)
Infected plants exhibit severe mosaic and stunting. Leaf distortion may occur and numerous side shoots produced.

Tan Leaf Spot (*Curvularia eragrostidis*)
Appears first as small green spots on the undersides of leaves. Spots coalesce and form tan, sunken, irregular-shaped spots. Leaves may become distorted.

SAINTPAULIA

Botrytis Blight (*Botrytis cinerea*)
See *Aphelandra*.

Corynespora Leaf Spot (*Corynespora cassiicola*)
See *Zebrina*.

Erwinia Blight (*Erwinia chrysanthemi*)
Blight of the crown, petioles, and leaves.

Powdery Mildew (*Oidium* sp.)
See *Kalanchoe*.

SANSEVIERIA

Bacterial Blight (*Erwinia chrysanthemi*)
Erwinia produces a brown to black root and crown rot. Leaf spots will occur from infection through the petioles, with the leaves and petioles ultimately becoming a greasy brown to black. Wilt and collapse of diseased plants is common.

Leaf Spot (*Fusarium moniliformae*)
Sunken, reddish-brown spots with yellowish borders up to $\frac{1}{2}$ in. in diameter may appear on one side of the leaf or may extend through it. The spots usually dry and fall out. If severe, numerous spots may run together, encircling the leaf and causing the part above to die.

Rhizome Rot (*Aspergillus niger*)
A rapid soft rot of stems which progresses into leaves under high temperatures. Lesions become sunken and water-soaked, eventually turning dark brown. Black, sooty masses of spores appear at the bases of lesions.

Southern Blight (*Sclerotium rolfsii*)
See *Brassaia*.

SCHEFFLERA

Alternaria Leaf Spot (*Alternaria panax*)
See *Brassaia*.

Bacterial Leaf Spot (*Erwinia chrysanthemi, Pseudomonas cichorii, Xanthomonas campestris* var. *hederae*)

Large, black spots 1 cm wide develop on leaf margins, tips, and petioles. Leaf drop is common.

Cercospora Leaf Spot (*Cercospora* sp.)

See *Brassaia*.

Rhizoctonia Aerial Blight (*Rhizoctonia solani*)

See *Fittonia*.

Southern Blight (*Sclerotium rolfsii*)

See *Brassaia*.

SINNINGIA

Bacterial Leaf Spot (*Pseudomonas* sp.)

See *Dracaena*.

Botrytis Blight (*Botrytis cinerea*)

See *Aphelandra*.

Myrothecium Crown Rot (*Myrothecium roridum*)

See *Aglaonema*.

SPATHIPHYLLUM

Cylindrocladium Root and Petiole Rot (*Cylindrocladium spathiphylii*)

Early symptoms include slight wilting and chlorosis of lower leaves. Leaves turn necrotic, petiole bases rot and detach from the plant.

Mosaic (Dasheen mosaic virus)

See *Aglaonema*.

Myrothecium Leaf Spot (*Myrothecium roridum*)

See *Aphelandra*.

Phytophthora Leaf Spot (*Phytophthora* sp.)

Irregularly shaped, black or brown spots up to 1 in. wide on leaf edges or centers. Spots are mostly near the base of infected plants.

Rhizoctonia Aerial Blight (*Rhizoctonia solani*)

See *Fittonia*.

Southern Blight (*Sclerotium rolfsii*)

See *Brassaia*.

SYNGONIUM

Bacterial Leaf Blight (*Xanthomonas dieffenbachiae*; Figure B. 16)

Figure B.16 Bacterial leaf blight of *Syngonium*. (Courtesy of Foliage Education and Research Foundation, Inc, Apopka, FL).

Water-soaked lesions occur along the leaf margin and leaf tip, and may elongate and extend into the midrib of the leaf. Initially dark green, the lesions turn yellow and eventually become brown and necrotic. A bright yellow zone often surrounds the diseased area. White flakes of dried bacterial cells are often visible on the lower surface of leaves having older lesions. The disease abounds in hot, humid environments, particularly where plants are crowded and overhead irrigation is used.

Black Cane Rot (*Ceratocystis fimbriata*)
    A black, water-soaked area sometimes girdling the stem. Leaves become chlorotic and die.

Colletotrichum Leaf Spot (*Colletotrichum* sp.)
    See *Aspidistra*.

Erwinia Blight (*Erwinia chrysanthemi*; Figure B.17)
    Any nonflowering, aboveground part may be infected. Oblong leafspots, often diffuse and irregular, vary from less than 1 mm to coalescing spots that may encompass most of the blade. Spots are blackish, often with a chlorotic margin.
    On the undersurface, lesions contain small, viscous drips from cream to tan to reddish brown in color. As the lesion dries, it becomes dark brown, often with a reddish-brown border surrounded by a chlorotic margin. Infected leaves are mushy, collapse, and may fall from the stem.

Leaf Spot (*Acremonium crotocinigenum*; Figure B.18)
    Irregular, water-soaked areas up to ⅛ in. in diameter occur on the leaf. With age, they may increase in size, and become reddish-brown with pale yellow borders. Centers become grayish and papery. The entire leaf may turn yellow and die if the infection is severe.

Figure B.17   *Erwinia* blight of *Syngonium*. (Courtesy of Foliage Education and Research Foundation, Inc., Apopka, FL).

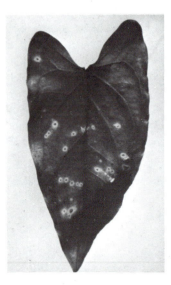

Figure B.18   *Acremonium* leaf spot of *Syngonium*. (Courtesy of Foliage Education and Research Foundation, Inc., Apopka, FL).

Myrothecium Leaf Spot *(Myrothecium roridum)*
   See *Aphelandra*.

Rhizoctonia Aerial Blight *(Rhizoctonia solani)*
   Lesions usually tan to black and appear water-soaked. When severe, the entire plant may become blighted and covered with the mycelium of the fungus.

Southern Blight *(Sclerotium rolfsii)*
   See *Brassaia*.

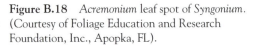

YUCCA

Gray Leaf Spot (*Cytosporina* sp.)

Spots on leaf tips and margins are generally gray with a brown border and may reach a size up to 7 cm. Lower leaves are most frequently affected.

Leaf Spot (*Coniothyrium concentricum*)

Light grayish-brown, zoned spots 1 in. or more in diameter with concentric rings of tiny black pycnidia appear on the leaves. Large portions of the leaf may be infected.

Southern Blight (*Sclerotium rolfsii*)

See *Brassaia*.

ZEBRINA

Bacterial Leaf Spot (*Pseudomonas woodsii*)

This disease appears as sunken, water-soaked spots with faded, reddish-yellow margins, becoming tan to black in the center with aging. *Setcreasea* and *Dichorisandra* are also susceptible.

Corynespora Leaf Spot (*Corynespora cassiicola*)

A wound-invading pathogen, *Corynespora*, produces dry, black, necrotic areas spreading to form a circular spot.

Mosaic (Tradescantia mosaic virus)

Infected plants exhibit severe mosaic and distortion of infected leaves. Plants are stunted.

Control of Leaf and Stem Diseases

Pathogenic stem and leaf diseases are not common indoors because of the low relative humidity. They do occur, however, and are often observed on newly acquired plants that were diseased when shipped by the producer. Hence, all new plants should be isolated and sprayed with suitable fungicides to eradicate existing diseases and prevent them from spreading.

In building interiors, preventive measures include the use of good cultural practices at all times, with particular attention to proper watering. Plants should not be overwatered, and the foliage should not be wet. Crowding of the plants should be avoided, and adequate ventilation provided. Destroying diseased leaves and/or plants will prevent further spread.

If a disease occurs, it should be diagnosed and preventive measures initiated at once. Should chemical control be necessary, use a recommended fungicide. Because disease-control materials and recommendations are constantly changing, it is best to consult the Agricultural Extension Service in your county or a reliable pesticide supplier for the current recommendations. For additional information about foliage plant diseases, consult *Compendium of Ornamental Foliage Plant Diseases* by A. R. Chase, listed in the references below.

REFERENCES

ALFIERI, S. A., JR.; *"Cercospora* and Edema of *Peperomia,"* Proceedings of the Florida State Horticultural Society 81: 388–391, 1968.

____, and J. F. KNAUSS: "Southern Blight of Schefflera," Proceedings of the Florida State Horticultural Society 83: 432–435, 1970.

____: "Stem and Leaf Rot of *Peperomia* Incited by *Sclerotium rolfsii,"* Proceedings of the Florida State Horticultural Society 85: 352–357, 1972.

ALFIERI, S. A., JR., and C. WEHLBURG: *"Cephalosporium* Leaf Spot of *Syngonium podophyllum* Schott.," Proceedings of the Florida State Horticultural Society 82: 366–368, 1969.

ANDERSON, R. G., and J. R. HARTMAN: *"Phomopis* Twig Blight on Weeping Fig Indoors: A Case Study," Foliage Digest 6(1): 5–7, 1983.

CHASE, A. R.: "Bacterial Leaf Spot of Crotons caused by *Xanthomonas,"* Foliage Digest 9(2): 1–3, 1986.

____: "Bacterial Leaf Spots of *Dracaena sanderana,"* Foliage Digest 7(10): 13–14, 1984.

____: "Chemical Control of *Myrothecium* Crown Rot and Leaf Spot of Rex Begonia," Florida Foliage 5(9): 14, 1982.

____: *Compendium of Ornamental Foliage Plant Diseases,* APS Press, The American Phytopathological Society, St. Paul, MN, 1987.

____: "Dasheen Mosaic Virus of *Dieffenbachias,"* Foliage Digest 6(5): 14, 1983.

____: "Diseases of Potted Foliage Plants—1990 Listing," Foliage Digest 23(11): 4–6, 1990.

____: "The Diseases That Damage *Pittosporum tobira,"* American Nurseryman 154(12): 13, 105–106, 1981.

____: *"Erwinia* spp.—Symptoms and Foliage Plant Hosts," Foliage Digest 6(7): 10–11, 1983.

____: "Indoor Plant Problems," Ornamental Outlook 2(6): 16, 1993.

____: "Leaf Spot of *Calatheas* Caused by *Alternaria,"* Foliage Digest 5(12): 5–6, 1982.

____: "Leaf Spot Diseases of Areca and Other Palms," Foliage Digest 5(10): 4–5, 1982.

____: "Three New Bacterial Diseases of Foliage Plants," Interior Landscape Industry 1(12): 48–50, 1984.

____: "*Xanthomonas* Diseases of Foliage Plants in Florida," *Foliage Digest* 10(2): 1–4, 1987.

____: "You Can Prevent Foliage Diseases," *Greenhouse Grower* 3(12): 36ff., 1985.

____, and C. A. CONOVER: "Rhizome Rot of *Sansevieria* spp. Caused by *Aspergillus niger*," *Foliage Digest* 4(7): 3, 1981.

____, and R. W. HENLEY: "Grape Ivy," *Foliage Digest* 8(8): 5–6, 1985.

CHASE, A. R., R. W. HENLEY, and L. S. OSBORNE: "*Maranta*," *Nurserymen's Digest* 19(9): 60–62, 1985.

CHASE, A. R., L. S. OSBORNE, and R. T. POOLE: "*Aphelandras*," *Nurserymen's Digest* 19(5): 22–24, 1985.

____: "Calathea," *Foliage Digest* 9(12): 1–4, 1986.

____: "*Syngonium*," *Foliage Digest* 7(11): 6–8, 1984.

CHASE, A. R., R. T. POOLE, and L. S. OSBORNE: "Areca Palm," *Foliage Digest* 8(10): 6–8, 1985.

____: "Bird's-Nest Fern," *Foliage Digest* 8(5): 3–5, 1985.

____: "*Dieffenbachias*," *Florida Foliage* 9(7): 47ff., 1983.

____: "Lipstick Plant," *Foliage Digest* 9(9): 4–6, 1986.

____, and R. J. HENNY: "*Spathiphyllum*," *Foliage Digest* 7(7): 6–8, 1984.

CONOVER, C. A., A. R. CHASE, and L. S. OSBORNE: "*Brassaia* and *Schefflera*," *Florida Foliage* 11(8): 50–51, 1985.

____: "Corn Plant," *Nurserymen's Digest* 19(5): 34–36, 1985.

CONOVER, C. A., L. S. OSBORNE, and A. R. CHASE: "Boston Ferns," *Nurserymen's Digest* 18(12): 37–38, 1984.

____: "Heart-Leaf Philodendron," *Foliage Digest* 7(6): 3–6, 1984.

HENNY, R. J., A. R. CHASE and L. S. OSBORNE: "Aglaonema," *Foliage Digest* 18(6): 1–4, 1995.

____, L. S. OSBORNE and A. R. CHASE: "Philodendron—Vining," *Foliage Digest* 18(5): 4–8, 1995.

KEIM, R., and R. G. MAIRE: "*Gliocladium* Disease of Palm," *Florida Foliage Grower* 14(9): 2–3, 1977.

KNAUSS, J. F.: *Common Diseases of Tropical Foliage Plants: I. Foliar Fungal Diseases*, ARC—Apopka Research Report RH-75-6, IFAS, University of Florida, Apopka, FL.

____: "Common Diseases of Tropical Foliage Plants: II. Bacterial Diseases," *Florists' Review* 153(3958): 27–28, 73–80, 1973.

____, and J. W. MILLER: "Description and Control of the Rapid Decay of *Scindapsus aureus* Incited by *Erwinia carotovora*," *Proceedings of the Florida State Horticultural Society* 85: 348–352, 1972.

KNAUSS, J. F., and C. WEHLBURG: "The Distribution and Phytogenicity of *Erwinia chrysanthemi* Buckholder et al. to *Syngonium podophyllum* Schott.," *Proceedings of the Florida State Horticultural Society* 82: 370–373, 1969.

KNAUSS, J. F., J. W. MILLER, and R. W. HENLEY: "Bacterial Blight of African Violet," *Florida Foliage Grower* 11(11): 4–5, 1974.

KNAUSS, J. F., J. W. MILLER, and R. J. VIRGONA: "Bacterial Blight of Fishtail Palm, A New Disease," *Foliage Digest* 1(11): 14–15, 1978.

MARLATT, R. B.: "*Aechmea fasciata* Has Serious Leaf Disease," *Foliage Digest* 2(4): 7, 1979.

____: "Brown Leaf Spot of *Dieffenbachia*," *Foliage Digest* 1(5): 10, 1978.

____: "Brown Leaf Spot of *Dieffenbachia*," *Plant Disease Reporter* 50(9): 687–689, 1966.

____: "A Serious Leaf Spot of *Ficus elastica* 'Decora' Stock Plants," *Foliage Digest* 2(8): 15–16, 1979.

MCRITCHIE, J. J., and J. W. MILLER: "*Corynespora* Leaf Spot of Zebra Plant," *Proceedings of the Florida State Horticultural Society* 86: 389–390, 1973.

____: "*Sclerotinia* Blight of *Gynura*," *Proceedings of the Florida State Horticultural Society* 87: 447–449, 1974.

MILLER, J. W.: "Bacterial Leaf Spot of Some *Commelinaceae* Caused by Carnation Leaf Spot Pathogen, *Pseudomonas woodsii*," *Proceedings of the Florida State Horticultural Society* 83: 435–437, 1970.

____, and C. WEHLBURG: "Bacterial Leaf Spot of *Dracaena sanderiana*," *Proceedings of the Florida State Horticultural Society* 82: 368–370, 1969.

MUNNECKE, D. E., and P. A. CHANDLER: "A Leaf Spot of *Philodendron* Related to Stomatal Exudation and to Temperature," *Phytopathology* 47(5): 299–303, 1957.

OSBORNE, L. S., A. R. CHASE, and D. G. BURCH: "*Chamaedorea* Palms," *Foliage Digest* 6(12): 1–4, 1983.

OSBORNE, L. S., A. R. CHASE, and C. A. CONOVER: "Croton," *Nurserymen's Digest* 18(6): 44–46, 1984.

OSBORNE, L. S., A. R. CHASE, and R. W. HENLEY: "English Ivy," *Nurserymen's Digest* 18(12): 71ff., 1984.

____: "Pilea," *Foliage Digest* 9(11): 3–6, 1986.

OSBORNE, L. S., C. A. CONOVER, AND A. R. CHASE: "*Cordyline* (Ti Plant)," *Foliage Digest* 8(9): 4–7, 1985.

____: "Norfolk Island Pine," *Foliage Digest* 9(6): 1–4, 1986.

OSBORNE, L. S., R. W. HENLEY, and A. R. CHASE: "Foliage Plant Research on *Ficus*," *Interscape* 5(40): 14–16, 18, 1983.

____: "Wax Plant," *Foliage Digest* 11(8): 1–4, 1986.

PIRONE, P. P.: *Diseases and Pests of Ornamental Plants*, 5th ed., John Wiley & Sons, Inc., New York, 1978.

POOLE, R. T., and A. J. PATE: "Bent-Tip of *Aglaonema commutatum mutatum* 'Fransher'" (abstract), *HortScience* 13(3): 389, 1978.

____: "Bent-Tip of *Aglaonema commutatum* 'Silver Queen'," *Foliage Digest* 3(8): 15–16, 1980.

POOLE, R. T., and A. R. CHASE: "Spathiphyllum Guide," *Greenhouse Grower* 5(5): 122–123, 1987.

____, and L. S. OSBORNE: "*Dracaena* 'Warneckii' and 'Janet Craig'," *Nurserymen's Digest* 19(6): 68, 70, 1985.

POOLE, R. T., L. S. OSBORNE, and A. R. CHASE: "Yucca," *Foliage Digest* 9(2): 3–5, 1986.

POWELL, C. C., JR.: "Planning Plant Protection," *Interior Landscape Industry* 8(3): 53–56, 1991.

RIDINGS, W. H., and J. J. MCRITCHIE: "*Phytophthora* Leaf Spot on *Philodendron oxycardium and Related Species*," *Proceedings of the Florida State Horticultural Society* 87: 442–447, 1974.

SEYMOUR, C. P.: "*Phyllosticta* Leaf Spot on *Dracaena*," *Florida Foliage Grower* 11(10): 4–5, 1974.

SIRADHANA, B. S., C. W. ELLETT, and A. F. SCHMITTHENNER: "Crown Rot of *Peperomia*," *Plant Disease Reporter* 52(3): 244, 1968.

TRUJILLO, E. E., A. M. ALVAREZ, and D. N. SWINDALE: "*Phytophthora* Leaf Spot of Ti," *Plant Disease Reporter* 59(5): 452–453, 1975.

WEHLBURG, C.: "Bacterial Leaf Spot and Tip Burn of *Philodendron oxycardium* caused by *Xanthomonas dieffenbachiae*," *Proceedings of the Florida State Horticultural Society* 81: 394–397, 1968.

WISLER, G. C., W. H. RIDINGS, and R. S. COX: "*Phytophthora* Leaf-spot of *Brassaia actinophylla*," *Proceedings of the Florida State Horticultural Society* 91: 240–242, 1978.

ZETTLER, F. W., W. H. RIDINGS, and R. D. HARTMAN: "Dasheen Mosaic Virus of Foliage Aroids," *Florida Foliage Grower* 11(7): 6, 1974.

ZETTLER, F. W., G. C. WISLER, and J. J. MCRITCHIE: "Dasheen Mosaic Virus of *Philodendron scandens* subsp. *oxycardium*," *Foliage Digest* 1(5): 15, 1978.

Index to Plants

General Index